Understanding DNA Ancestry

T0188317

DNA ancestry companies generate revenues in the region of $1 billion each year, and the company 23andMe is said to have sold 10 million DNA ancestry kits to date. Although evidently popular, the science behind how DNA ancestry tests work is mystifying and difficult for the general public to interpret and understand. In this accessible and engaging book, Sheldon Krimsky, a leading researcher, investigates the methods that different companies use for DNA ancestry testing. He also discusses what the tests are used for, from their application in criminal investigations to discovering missing relatives. With a lack of transparency from companies in sharing their data, absent validation of methods by independent scientists, and currently no agreed-upon standards of accuracy, this book also examines the ethical issues behind genetic genealogy testing, including concerns surrounding data privacy and security. It demystifies the art and science of DNA ancestry testing for the general reader.

Sheldon Krimsky is Lenore Stern Professor of Humanities and Social Sciences and Adjunct Professor of Public Health and Community Medicine at Tufts University, Massachusetts, USA. His research has focused on the linkages between science/technology, ethics/values, and public policy. His areas of specialization include biomedical sciences, bioethics, science and technology studies, risk assessment and communication, social history of science, and environmental health. He is the author of over 200 articles and reviews, and author, coauthor, or editor of 16 books.

The **Understanding Life** series is for anyone wanting an engaging and concise way into a key biological topic. Offering a multidisciplinary perspective, these accessible guides address common misconceptions and misunderstandings in a thoughtful way to help stimulate debate and encourage a more in-depth understanding. Written by leading thinkers in each field, these books are for anyone wanting an expert overview that will enable clearer thinking on each topic.

Series Editor: Kostas Kampourakis http://kampourakis.com

Published titles:

Understanding Evolution	Kostas Kampourakis	9781108746083
Understanding Coronavirus	Raul Rabadan	9781108826716
Understanding Development	Alessandro Minelli	9781108799232
Understanding Evo-Devo	Wallace Arthur	9781108819466
Understanding Genes	Kostas Kampourakis	9781108812825
Understanding DNA Ancestry	Sheldon Krimsky	9781108816038

Forthcoming:

Understanding Intelligence	Ken Richardson	9781108940368
Understanding Metaphors in the Life Sciences	Andrew S. Reynolds	9781108940498
Understanding Creationism	Glenn Branch	9781108927505
Understanding Species	John S. Wilkins	9781108987196
Understanding the Nature–Nurture Debate	Eric Turkheimer	9781108958165
Understanding How Science Explains the World	Kevin McCain	9781108995504
Understanding Cancer	Robin Hesketh	9781009005999
Understanding Forensic DNA	Suzanne Bell and John Butler	9781009044011
Understanding Race	Rob DeSalle and Ian Tattersall	9781009055581
Understanding Fertility	Gab Kovacs	9781009054164

Understanding DNA Ancestry

SHELDON KRIMSKY
Tufts University, Massachusetts

CAMBRIDGE
UNIVERSITY PRESS

CAMBRIDGE
UNIVERSITY PRESS

University Printing House, Cambridge CB2 8BS, United Kingdom

One Liberty Plaza, 20th Floor, New York, NY 10006, USA

477 Williamstown Road, Port Melbourne, VIC 3207, Australia

314–321, 3rd Floor, Plot 3, Splendor Forum, Jasola District Centre,
New Delhi – 110025, India

103 Penang Road, #05–06/07, Visioncrest Commercial, Singapore 238467

Cambridge University Press is part of the University of Cambridge.

It furthers the University's mission by disseminating knowledge in the pursuit of
education, learning, and research at the highest international levels of excellence.

www.cambridge.org
Information on this title: www.cambridge.org/9781108841986
DOI: 10.1017/9781108895651

First published 2022

Printed in the United Kingdom by TJ Books Limited, Padstow Cornwall

A catalogue record for this publication is available from the British Library.

ISBN 978-1-108-84198-6 Hardback
ISBN 978-1-108-81603-8 Paperback

"Sheldon's book represents a much needed historical, technical, and ethical treatment of this rapidly evolving and growing industry. *Understanding DNA Ancestry* tackles a complex topic that many are fascinated by but few have the educational background to appreciate fully un-shepherded, and does so in a way that is accessible and easy to internalize by the very lay readers who have literally built the entire industry with their demand. His book is not only timely, but way, way overdue, and is special for many reasons. For one, this book comes from an author who is independent of the industry; the vast majority of white-paper, journal publications, editorials, and books do not, and often come with unfair biases and agendas. Secondly, and uniquely, the book covers the entire spectrum of societal, technical, and potentially personal implications of this new technology. With comprehensive scope, Sheldon brings you to appreciate fully the mechanics of the methods; the impact of this new technology on medicine, privacy, and other ethical considerations; their upsides and downsides, power and limitations; and how they may affect you personally – we come to appreciate the technology in its universal totality, including even its impact on Forensic Science, in terms not just of phenotyping (ascribing physical characteristics from crime-scene DNA), but of genetic genealogy, which is the ultimate evolution of the technology that almost all ignore, but that touches us all in terms of public safety. Sheldon's book is historically and technologically accurate, socially responsible, objective, inclusive in scope, and sensitive to personal risks and rewards. As such, this book is in my view instrumental for anyone considering a genetic ancestry test. If you are a lay consumer of genetic ancestry testing products, it has my highest recommendation for you."

Tony N. Frudakis, Ph.D., Forensic Scientist, Albuquerque Police Department DNA Laboratory, and Founder DNAPrint Genomics, Inc. (1999)

"This book has it all – science and technology, history, ethics, law, and interesting stories of genealogy. It is classic Krimsky – a truly scholarly endeavor made incredibly approachable. Krimsky goes into sufficient depth to empower the reader with the background necessary to appreciate and understand DNA ancestry fully. The book is comprehensive, describing the key discoveries leading to the modern science of ancestry, including the history and development of the multiple generations of technologies used to achieve the resolution of understanding we have today. Applications of the technology's uses and misuses are covered, as well as privacy and ethical considerations. Krimsky is a terrific storyteller of individual cases, where people found out they weren't who they thought they were. He remains in the background throughout as a balanced and unbiased observer. A most interesting and timely book that will inform, entertain, and empower the millions who have had or are considering a consumer DNA test."

David R. Walt, Harvard Medical School

"How do private companies, like Ancestry.com and 23&me, use DNA to determine your ancestry? How does this differ from forensic DNA used by police and the FBI? And should we be worried about some or all of this? In this compelling book, Sheldon Krimsky provides clear, informative, and nuanced answers to all these questions and more."

Naomi Oreskes, Henry Charles Lea Professor of the
History of Science and Affiliated Professor of Earth
and Planetary Sciences, Harvard University

"Professor Sheldon Krimsky's book *Understanding DNA Ancestry* introduces readers to the vast panoply of complications that can face those who are interested in the sources of their ancestors. Such readers may be surprised that results from different genetic ancestry companies can vary in their conclusions about the results from company to company. Test-takers may sometimes confront unexpected surprises they had never suspected about co-existing family members, or discover that an assumed relationship may not be genetically based. They will also learn that some ancestry companies help the FBI in identifying criminals using those companies' DNA sequencing collections. What is more, the eager readers may find out that they are learning more than they knew about genetics. Entering these new studies raises questions about whether supposed classical races of peoples have a genetic basis at all.

Professor Krimsky has a long career in preparing detailed collections that have been important to fields of science."

Jonathan Beckwith, Professor Emeritus, Harvard Medical School, Department of Microbiology (and a geneticist)

"In the last three years, there has been a sharp surge in genetic ancestry testing, not only in the US, but across the globe. There are now more than seventy companies promoting an array of such offerings, which range from recreational to medical to forensic uses and claims. Sheldon Krimsky has provided an illuminating social history of these developments, with lucid prose that explains the uses and limits of such testing. But *caveat emptor*: many consumers will be dismayed to learn that the seductive lure and broad claims commonly outstrip the capacity of these tests to provide clear and replicable results."

Troy Duster, Chancellor's Professor Emeritus, University of California, Berkeley

"Sheldon Krimsky has written an illuminating description in amazingly simple language on a most difficult but essential subject – ancestry.

Krimsky's explanations of how DNA ancestry works and what it means for modern society are essential contributions to how we, as humans, understand our own variation. Understanding these important aspects of our variation is critical to our world-view and the place of our species in the modern world."

Robert DeSalle, American Museum of Natural History,
New York

For Eliot Krimsky, Lisa Benger, and Siona Rose Krimsky

Contents

Foreword

Who am I? Where do I come from? The answers to these questions might seem self-evident to many people. Being aware of the history of their ancestors, their genealogical ancestry, they are able to talk about their ancestry, their ultimate "roots," and their resulting ethnicity/nationality. Others are still asking these questions, perhaps because they lack information about their genealogical ancestry. To find answers, they often turn to DNA tests that are supposed to provide information about their genetic ancestry. Thus, people hope, based on their presumed genetic ancestry, to be able to make inferences about their genealogical ancestry. However, this is not as simple as it may seem. Currently, millions of people have taken DNA ancestry tests, some hoping to find such answers, others just doing it as a recreational activity. But all these seem to take the validity and the utility of these tests for granted, often overlooking the fact that these tests are based on specific assumptions and have important limitations. The present book by Sheldon Krimsky does a lot at the same time: It diligently explains the science of DNA ancestry testing, it responsibly considers its limitations and pitfalls, and it sensitively discusses the related ethical and legal issues. If you are thinking about taking a DNA ancestry test, or if you are simply interested in this topic, then this fascinating book will guide you through the various issues at stake. It is a tour-de-force that provides an informed and pragmatic view of DNA ancestry testing that would be useful to anyone interested in really understanding this topic.

Kostas Kampourakis, Series Editor

Acknowledgments

I wish to thank David Walt, Tony Frudakis, and Michael Carson for reading selected chapters and providing feedback, John M. Coffin for clarifying some historical points about the first bacteria and viruses fully sequenced, and Bennett Greenspan, CEO of Family Tree DNA for agreeing to an interview. I am indebted to the excellent editors at Cambridge University Press: Kostas Kampourakis and Katrina Halliday.

1 Introduction

Stories of family deceits and deceptions have become commonplace in a media receptive to personal tales of triumph and tragedy. A distinguished geneticist learns in his mature years that his mother, while married to his legal father, had a secret affair that begat him. A best-selling author discovers that her paternal DNA was from a medical student serving as a sperm donor and not her legal father, who traced her ancestry deep into Eastern Europe. A woman who, as a newborn, was left in a bag abandoned in the foyer of a Brooklyn apartment building searches for her biological parents 23 years later. These revelations are the result of the millennial DNA ancestry revolution.

I first learned about the surge of public interest in DNA genealogy when I was researching a book on forensic applications of DNA. After the British scientist Alec Jeffreys created a way to use DNA sequences to provide unique identifiers of individuals in 1983, which later became known as DNA fingerprinting, police agencies throughout the world adopted the DNA identification system. In the United States, the forensic DNA system of personal identification was the basis of a databank known as CODIS (Combined DNA Index System), established in 1990. The US Federal Bureau of Investigation (FBI) created CODIS as a pilot project involving 14 states and local laboratories before it became national. The FBI has a database containing over 12 million DNA profiles and similar databases exist worldwide. The early techniques of gene sequencing used by the FBI were based on markers called short tandem repeats, or STRs (see Chapter 5), which provided no information about a person's appearance or genealogy.

Familial and Historical Genealogy

The tools of genetic sequencing were turned into a commercial opportunity when new genetic markers were developed and the sequencing methods improved. Two strands of interest among people active in exploring genome sequencing were genealogy and health. Those interested in familial genealogy wish to learn about their family ancestry over a few generations. Those interested in historical genealogy want to find genetic connections over hundreds or even thousands of years, utilizing all the tools of historical and genetic research. Someone may, for example, be interested in whether their family tree can be traced to the Roman Empire.

Medical genome sequencing, which predated genealogical sequencing, searches for specific mutations that are linked to diseases. The markers sequenced on the genome are much more specific than genealogical sequencing.

Traditionally, genealogical research – whether familial or historical – has largely been considered a hobby or a family recreational activity. The techniques required for familial genealogy include archival research involving sources such as the federal census, birth and death certificates, probate records, gravestones, obituaries, tax records, church and military documents, immigration records, and all the tangible physical records a person leaves behind when they are gone. With the digitization of documents, these searches have become easier and available to more people.

The Birth of Genome Sequencing

Sequencing the DNA of an organism took off in the 1980s in the fields of medicine and criminal justice after Fred Sanger, Walter Gilbert, and Paul Berg shared the Nobel Prize for their development of DNA sequencing methods. The first bacterial virus (bacteriophage) sequenced in its entirety was reported in 1977. The first animal virus completely sequenced was SV40 in 1978. These viruses had small genomes, were easy to grow and purify, and had been studied as models for many years, which explains why they were sequenced first. The first human virus fully sequenced was the polio virus, in 1981.

DNA sequencing and genetic databases also became tools of anthropologists and population geneticists, who used them to study ancient population movements. By analyzing specific DNA changes (called polymorphisms) in groups of people, scientists were able to track migrations of populations across the globe.

The commercial DNA ancestry sector began in 2000 with the launch of Family Tree DNA, the first of dozens of companies to capture public interest in genealogy. While he was investigating his European ancestry, Bennett Greenspan, the company's founder and CEO, learned that scientists were able to use DNA to track a person's ancestry. He understood there was a void in the marketplace for such applications in genealogical searches. Within a couple of decades, the commercial ancestry sector was occupied by around 70 companies. In order to apply the most current and sophisticated methods of genomic genealogy, companies hired geneticists, anthropologists, statisticians, and experts in data analytics.

Goal for the Book

As I immersed myself in the scientific literature and popular books and articles about DNA ancestry, I came to realize that it took more than a basic understanding of genetics to understand how one's ancestry could be read from one's genome. It was a significant departure from traditional genealogical methods which involve the construction of family trees encompassing several past generations. In contrast, DNA ancestry can take one back many generations to people whose names are not known but who occupied certain regions of the world and share some segments of DNA with the customer. Where traditional genealogical methods involving the search of family documents is straightforward, DNA ancestry is complex and requires investment in equipment and expertise.

When I began looking into the stages involved in developing an ancestry DNA profile of an individual, I realized I was looking into black boxes that I could barely decipher. Forensic DNA methods have a universally accepted and validated methodology because the technology is simpler and was funded by the government, but the same cannot be said for company-developed ancestry DNA profiles. The literature on DNA ancestry tests was bifurcated between technical scientific studies and popular magazine and newspaper articles. The scientific studies are out of reach for most readers and the popular stories are deficient in explaining the science.

My goal in writing this book is to decipher the process of DNA ancestry testing and to demystify the elusive technical components while exploring the applications of genealogical ancestry beyond that of creating family networks. To begin, I recognized that there are a number of terms that appear in the scientific, commercial, and popular writings about the role of DNA in understanding biogeographical populations (groups of people with distinct genetic markers inhabiting regions of the world) and their descendants. Terms such as ancestry, genealogy, pedigree, descent, lineage, and genetic inheritance are among the most frequently used.

The Varieties of Ancestry

In an essay titled "What is ancestry?" geneticists Iain Mathieson and Aylwyn Scally made some important distinctions such as that between genealogical ancestry and genetic ancestry. Genealogical ancestry refers to identifiable ancestors in your family tree, which is constructed from historical documents such as those in public records, as well as family lore. If you search back n generations, you will have 2^n ancestors. For example, your great-grandparents are three generations away from you, and your family tree exhibits $2^3 = 8$ great-grandparents. The term "pedigree" refers to how all your genealogical ancestors are related to one another: great-grandfather, second cousin, third cousin once removed, etc. Genealogical ancestry has its limitations because few people have comprehensive knowledge of their ancestors beyond a small number of generations, for which they may not even have records. Assuming a generation of 25 years, on average, 250 years or 10 generations ago there existed, in theory, 2^{10} or 1,024 ancestors for each one of us.

Genetic ancestry refers to the people in your past who contributed to the composition of your genome beyond the 50 percent from your parents and 25 percent from your grandparents. For many generations in the past, the genealogical family will have many people who do not share genetic sequences because a descendant inherited the DNA sequences of some ancestors but not those of others. Therefore, a person's genetic ancestry consists of a small part of one's genealogical ancestry. One way to understand the differences between genealogical and genetic ancestry is that full siblings have identical genealogical ancestry but differ in their genetic ancestry, because they inherit different chromosomal segments from their parents. Two siblings have the same parents but have not inherited exactly the same DNA from them. Rather, two siblings have, on average, 50 percent of the same DNA as each other.

Traditional genealogical research seeks to construct family trees of individuals, exhibiting the pedigrees (relations between your genealogical ancestors) of descent. Genetic ancestry allows scientists to compare individual genomes with the average genome of a reference sample of some population, which may not be a random or representative sample, but a convenience sample that the ancestry company was able to obtain. While you might have no ancestral link with most members of the population, you might share certain population-specific markers indicating the region of your descent. Thus, genetic ancestry has led to population ancestry. What geneticists and ancestry DNA companies mean by ancestry is the genetic similarity between individuals and populations.

Cultural ancestry is another category of how people relate to their genealogy. Native American ancestry is based on whether someone was embedded in the cultural traditions of a tribe, not on their DNA or on the construction of a family tree. Native American cultural studies scholar Kim TallBear argues in her book *Native American DNA* that DNA connections are no substitute for kinship relations. Much of kinship and tribal citizenship is biological, but not in ways captured by genetic lineages.

Consider a person with a single Native American great-great-great-grandparent. They might not have inherited any Native American chromosome segments, so their genetic ancestry would be 0 percent Native American. Yet, if they were brought up in a Native American tribe, that is their cultural ancestry.

The practice of genealogy has been popularized by television series such as Britain's *Who Do You Think You Are?* and the US Public Broadcasting System series *Finding Your Roots*. These programs have contributed to the commercial market for genealogy. What are these programs tapping into? Cultural studies scholar Julie Rak's answer to the question of why so many people around the world have become interested in genealogy is that "doing genealogy is about 'doing kinship,' a way to facilitate connection between the living and the dead, to construct identity not just from one's own experience but from knowledge of one's ancestors, to work through grief and loss."

Structure of the Book

The book is organized around 11 thematic chapters. Chapter 2 discusses the business behind DNA ancestry and how gathering genomic data has allowed companies to create a bifurcated business model of genealogy and health, where the same company collects both types of information. Chapter 3 explores the origin of early populations and discusses the "Out of Africa" theory and its significance in understanding genetic diversity in human populations. I begin to explore the science behind DNA ancestry testing in Chapter 4, which introduces the reader to the concept of genetic markers. Some of the early patents submitted for DNA ancestry tests provide unusual clarity about the science used in the process. While not all the patent applications were awarded, they give us an insight into how early inventors conceptualized the methods of using DNA for disease risk diagnosis, for reading phenotypes, and for ancestry inference.

Chapter 5 examines different markers used in ancestry testing from mono-parental DNA markers (markers that appear exclusively on either the male or female genome), such as on the Y chromosome of the male and the female mitochondrial DNA, to bi-parental markers (which appear on both male and female genomes) situated on the autosomes (all the chromosomes except the sex chromosomes). In Chapter 6, I discuss the reference panels used by DNA ancestry companies. Markers on the consumer DNA sample are compared to markers on reference panels, which are supposed to represent populations in different regions of the world. These reference panels tend to be proprietary for each company, although they may in part utilize public data. Chapter 7 examines how a person's DNA is compared to population reference panels.

A critical component in DNA ancestry testing is reading the thousands of markers on the donor's DNA sample. Chapter 8 discusses the development of the microarray for analyzing the DNA of an ancestry customer and identifying the DNA markers in the sample. With the development of the microarray, the cost of analyzing a DNA sample was dramatically reduced, making it possible to charge approximately $100 for an estimate of ancestry.

The growth of the DNA ancestry sector has also had repercussions in criminal justice and provided a new method of finding criminals from crime-scene evidence when DNA matches on national databases did not reveal suspects. Chapter 9 discusses new police methods for finding felons through open-access DNA databases.

Chapter 10 explores some of the ethical and privacy issues in commercial ancestry tests, including whether they reinscribe race into the scientific vernacular and reinforce genetic essentialism, and how one person's DNA ancestry test affects other members of one's family. It also discusses why, for some people, the results of their DNA ancestry tests provide validation and assurance of their personal identity. This chapter explores how people use the DNA ancestry tests to establish their ethnicity and when this genealogical information is advantageous to them on social or civil rights grounds.

Many people are now known to have discovered unknown relatives or mistaken relatives through their DNA ancestry results. Chapter 11 selects some iconic stories to illustrate the elation and heartbreak of discovering the truth about family. Chapter 12 examines the accuracy, consistency, and validation of DNA ancestry testing. It explores why test results among different companies vary and why twins tested by the same company have given different results. Chapter 13 concludes the book.

2 The Business of DNA Ancestry

Over a period of 20 years, family genetic genealogy, through the purchase of consumer ancestry testing kits, has been one of the fastest growing family activities of this generation. Citing data from the International Society of Genetic Genealogy, the *Washington Post* reported in 2017 that 8 million people worldwide were involved with recreational genomics. It is estimated that by 2019 about 25 million people had signed up for a DNA ancestry test offered by one of the dozens of companies that have entered this marketplace. The kits are sent to a person's home with return packaging that includes a reservoir for depositing saliva or swabs for sampling cheek cells. The *MIT Technology Review* predicted that by 2021 there would be 100 million consumers of ancestry DNA services.

In 2010, researchers reported 38 companies worldwide had entered the home DNA ancestry marketplace. Six years later, 74 such companies were competing, and by the end of 2019, 61 ancestry testing companies were identified by the International Society of Genetic Genealogy. Some of the earlier companies were either absorbed by competitors or disappeared.

The First DNA Ancestry Company

The first company offering direct to consumer (DTC) genetic ancestry DNA tests was Family Tree DNA (FTD), which was incorporated in the year 2000. Initially, it used mitochondrial DNA and Y chromosome DNA for ancestry testing. DNA Print Genomics offered the first genomic (autosomal) ancestry test in 2002, which soon became the standard for other companies. Autosomal tests utilize the entire human genome.

After several years, FTD partnered with National Geographic, which had founded the Genographic Project in 2005, the goal of which was to collect DNA samples throughout the globe to understand the patterns of human migration. By 2019, the Genographic Project claimed over 1 million participants in 140 countries. FTD needed the worldwide DNA samples to advance its commercial venture for paid consumer ancestry testing.

FTD was founded by Bennett Greenspan, who had a childhood interest in genealogy. In seeking to trace his ancestors from Poland, he learned that some had emigrated to Buenos Aires, Argentina. After reading that Thomas Jefferson's descendants could be traced by DNA, he contacted geneticist Michael Hammer, a professor at the University of Arizona whose name had been on a DNA ancestry publication known to Greenspan. That's when he realized there was a business opportunity in using DNA to trace ancestry. After selling his photographic supply business he turned his avocation interest in genealogy into a commercial venture.

Greenspan collected DNA samples from people in various regions of the world and used existing publicly funded, open-access DNA databases of different ethnicities. He signed up a group of Sephardic Jews from Seattle, Washington, for DNA tests that allowed him to build a Sephardic DNA database. This was a database of opportunity built through Greenspan's contacts. After purchasing some smaller genetic testing companies, in 2011 FTD renamed itself Gene By Gene, and within eight years had a staff of 150 people.

Growth of the Digital Ancestry Sector

Of the 246 DTC companies listed in 2018, 74 offered ancestry services. Others focused on legal paternity, maternity, grandparent identification, and sibling identification testing.

In 2010, the DNA ancestry industry was valued at $15 million; and six years later, in 2016, that value had grown to $173 million. The two giant ancestry DNA companies, by virtue of the number of tests kits sold, are AncestryDNA, founded in 2007 as an offshoot of National Geographic, located in Lehi, Utah, and 23andMe, founded in 2006 by Linda Avey and Anne Wojcicki (funded by Google), located at Mountain View, California. Three years after incorporation, AncestryDNA had acquired 1 million customers. It started a health division in 2015 and began collecting customer health information. By November 2019, AncestryDNA reported that it had distributed 14 million kits. Meanwhile, 23andMe established a partnership with GlaxoSmithKline to leverage genetic data for drug development. The business model for a number of companies, including 23andMe, involved selling the DNA data to pharmaceutical companies. Drug development benefited from the large genome databases that could provide information on which mutations were antagonistic to specific drug therapies and how rare these mutations were. Figure 2.1 shows the rapid growth of autosomal tests administered by five ancestry companies between 2012 and 2020.

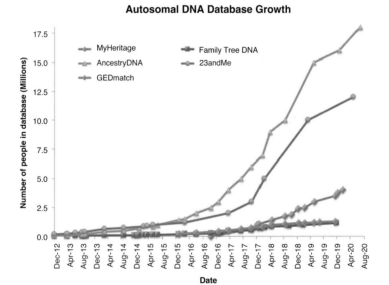

Autosomal DNA Database Growth

Figure 2.1   Growth in autosomal DNA ancestry testing. Source: Leah Larkin, 2020. Autosomal DNA testing growth. https://thednageek.com/dna-tests.

Ancestry and Health

Health is an additional market area of expansion for DTC ancestry tests. There were already companies that specialized in testing for DNA health indicators such as breast cancer mutations. When adding health indicators to their consumer DNA ancestry tests, ancestry companies expected that consumers would take the information to their physicians. In 2019, AncestryDNA gave a million-dollar educational grant to the company UpToDate, founded in 1992, whose mission is to provide credible medical information to physicians, including genomic information from consumers of DTC ancestry tests.

There are two reasons why DNA ancestry companies want to expand into the consumer health market. First, they can add a premium charge to the ancestry test by offering health indicators in the genome of the consumer. Second, the health genomic data they accumulate can be sold to pharmaceutical companies and medical research institutions. Some companies only provide genomic health indicators – that is, mutations that provide information about the risk of being afflicted by a particular disease.

The term "two-sided market strategy" has been used to describe the DTC genetic testing sector. A two-sided market occurs when two different user groups interact through an intermediary economic platform, known as a two-sided platform. This set up makes possible exchanges that would not otherwise have occurred, creating value for both sides – in this case, the people seeking DNA analyses and the institutions seeking to obtain information about them, who can both be considered consumers. Some DNA ancestry companies give consumers the option of allowing their DNA to be used in research.

A chief science officer of one company noted that DTC services have turned their genome collections into cash cows. For example, 23andMe has dozens of partnerships with drug companies for access to part or all of their collections. While AncestryDNA expands its capability to collect DNA health information, it needs to assess whether there are markets that will benefit from such information.

As genomic ancestry companies providing heritage information have expanded into healthcare, they have drawn criticism from the medical community. The privacy of medical data is, after all, protected by law in the United States, and this sets a higher bar of responsibility for DTC testing companies. As they expand their disclosures, they straddle the line between consumer focus and medical practice. Because European countries have national healthcare systems, the genetic privacy issues prevalent in the United States, especially with regard to healthcare discrimination, were not an issue in Europe.

The CEO of 23andMe, Anne Wojcicki, has been quoted as saying: "What the medical world thinks people want or should want, or should have, is not reflective of what consumer interests actually are." In 2008, *Time Magazine* named the retail DNA test by 23andMe as its Invention of the Year.

As noted by University of Pennsylvania legal scholar Jennifer Wagner, some critics of the DNA ancestry testing industry claim that companies are selling medical information, and thus practicing medicine without a license. The line between providing information on genetic risk factors and practicing medicine is blurry. Companies like 23andMe argue that people have a right to their health information at a reasonable price, but in providing that information they deny that they are practicing medicine. Nevertheless, health agencies feel that the public deserve validated health tests.

In 2010, the US Food and Drug Administration (FDA) stopped the health DNA services offered by 23andMe for not applying to the FDA for approval of its tests. The agency was concerned about the public health consequences of inaccurate tests from 23andMe's Personal Genome Service, and interpreted this service as a medical device that needed approval. The company had to withdraw a couple of hundred medical tests in 2013 until they gained approval for each test.

In 2017, 23andMe obtained FDA approval to provide test information for two breast cancer genes, and soon after was given approval to test for prostate cancer and Parkinson's disease. Such mutations, called disease markers, are risk factors and do not provide definitive evidence that a carrier will suffer either disease.

There has been no US government oversight of DNA ancestry services. Deakin University (Australia) scholars Elizabeth Watt and Emma Kowal noted in 2019 that this lack of oversight supports the idea that genetic searches of one's ancestry is a relatively harmless hobby. But with advancements in genomic science and the history of population movements, DNA ancestry has become a credible source of genealogical information. As *Nature* reported, "Commercial ancestry testing, once the province of limited information of dubious accuracy, is taking advantage of whole-genome scans, sophisticated analyses and ever-deeper databases of human genetic diversity to help people answer a simple question: where am I from?"

Expanding the Types of Genetic Tests

Even though ancestry testing, in its intended sense, does not involve health, some critics writing in *Science* in 2009 stated that there should be oversight for false claims, if nothing else. They argued that genetic ancestry tests fall into an unregulated no-man's land with little oversight and few industry guidelines to ensure the quality, validity, and interpretation of the information sold.

Some DNA ancestry companies provide incentives to customers by allowing them to download the raw genome data that the company generates from their saliva or cheek swab submissions. Nebula Genomics awarded tokens to customers who provided phenotypic data (information about their physical characteristics such as weight, height, and pre-existing health conditions), and with enough tokens they can purchase their whole genome sequence. Customers would have access to the single nucleotide polymorphisms (SNPs) – single "letter" changes in a DNA sequence – detected by the company's microarray chips, which can reveal thousands of sequences at locations in the genome. That data can be uploaded on open-access platforms that provide a service for comparing a contributor's DNA with that of other people who have similarly uploaded their raw data. In this way, people can learn about possible relatives about whom they were unaware. Ancestry companies that do not have open-access platforms can only provide matches of potential family members of people who, because they have purchased a particular company's test, are on their own databases.

In mid-2000s, companies began aggressively advertising their services to position themselves against their competitors. When AncestryDNA began in 2007, it offered customers testing for mitochondrial DNA (inherited from mother to offspring) and Y chromosome DNA (inherited from father to son), which could provide genealogical information about maternal or paternal lineages. In 2012, it introduced a DNA test utilizing full genomic DNA (autosomal) sites of over 700,000 marker locations to determine ethnic backgrounds and to trace distant cousins. The company announced a test that included 22 geographic and ethnic categories, including six regions in Europe, five regions in Africa, and Native America. In 2016, AncestryDNA spent $109 million on TV and other ads. By 2017, purchases of AncestryDNA tests doubled, exceeding 12 million, attributed largely to the intensive advertising.

In 2019, when it was reported that 74 companies offered ancestry DNA tests, DTC genetic testing was considered a billion-dollar industry. On May 31, 2019, AncestryDNA reported selling its service to more than 15 million customers.

Behind Consumer Interest in Ancestry Tests

Why do consumers research their ancestry? This was the question asked by Carolyn Strong and her collaborators at Cardiff University of Wales and Queensland University of Technology in Brisbane, Australia, in their paper published in the *Journal of Business Research* in 2019.

The authors began their study with a hypothesis that for many customers of DNA ancestry tests, "the desire for genealogical products is driven by uncertainty about their sense of self." Their research was informed by the *uncertainty-identity theory*, which stresses that people who exhibit self-uncertainty are motivated to reduce this uncertainty, and a way to do this is identifying with cohesive groups, which allows them to feel more certain of the world. Thus, DNA ancestry provides a vehicle for people to connect with historical groups who share similar segments of DNA.

The research team surveyed 226 participants, who were asked to read advertising messages about a fictitious DNA testing service. Using a seven-point scale from *strongly disagree* to *strongly agree*, the participants were asked questions that were designed to elicit their level of identity uncertainty and their desire to reduce that uncertainty. The investigators found that participants who were uncertain about their identity and had a high need for closure about that uncertainty tended to evaluate genealogical products favorably. They concluded that people with a high need for closure are more likely to sign up for DNA testing where uncertainty reduction is mentioned.

Other researchers have concluded that personal curiosity about one's ancestral roots is only one of several reasons that people undertake such tests. They also engage in ancestry testing for political and economic reasons – by proving their membership in a tribe, for example, they might qualify for health and financial benefits, or minority heritage might qualify them for diversity quotas at elite universities.

With all the hoopla around DNA ancestry tests, they became ideal family gifts for stimulating discussions about test outcomes and for engaging children in family genealogy pursuits. Despite the decade-long rise in sales, in 2020 there was a sudden decline in interest. Two of the leading companies, 23andMe and AncestryDNA, experienced declines in sales of DNA ancestry kits of 54 and 38 percent, respectively. The decline was attributed to market saturation, economic recession related to the COVID-19 pandemic, and privacy concerns.

Julia Creet, professor of English at York University in Toronto wrote: "Almost every database shares information with the pharmaceutical industry. 23andMe was clear from the beginning that its health information would be used by its research partners and asked consumers to consent. But when it started to sign major deals with drug developers in 2015, consumers began to realize ... they were its product." Some market analysts contend that consumer interest in privacy, especially after the highly publicized story about how police caught the Golden State Killer in 2019 from ancestry DNA data, brought genetic privacy into the public mind, which was the major factor in turning people away from getting tested (see Chapter 9).

Another possible factor in the reduction in sales is the publicity around the validity of tests, which has been highlighted on social media, citing disparate results from twin ancestry tests. Because the methods for undertaking ancestry DNA genealogical analysis are not standardized, in contrast to forensic DNA identification, and because companies have their own unpublished and proprietary reference databases, it is not unusual for the outcomes of one's genetic ancestry test to vary across companies. More about the validation and consistency of DNA ancestry tests can be found in Chapter 12.

In its 2008 report on DNA ancestry tests, the American Society of Human Genetics (ASHG) emphasized that the methods of companies offering ancestry tests left a lot to be desired. They provide incomplete representation of human genetic ancestry in their genetic databases, hold false assumptions that contemporary population samples provide reliable information about ancestral populations, and lack transparency on the statistical methods and algorithms companies use to determine results. According to the ASHG, without mechanisms to enforce transparency, the scientific basis for assertions of biogeographical ancestry cannot be assessed. Biogeographical ancestry is the component of ethnicity that is biologically determined and can be estimated by using genetic markers. Since the ASHG report cited 30 ancestry companies, many outside the United States, one may assume that its report and recommendations apply to all of them (see Chapter 12).

3 What Our Genomes Tell Us about the Geographical Origins and Movements of Early Human Populations

The purpose of DNA ancestry genealogy is to determine what the geographical origins are of an individual's ancestry, regardless of where he or she is currently living. The scientific premise behind this exercise is that people's DNA contains sequences of their ancestors' DNA, which can be traced back hundreds or even thousands of years, and that their ancestors were settled in a region of the world that remained relatively isolated. This isolation allowed ancient populations to remain inbred within certain geographical parameters. Inbreeding is the mating of humans closely related by ancestry. It is more likely to occur in isolated, non-migrating populations, resulting in a loss of genetic diversity and a high incidence of birth defects. Mutations in the DNA circulating within these inbred populations can provide a genetic fingerprint of the geographical region in which they were located.

In 10 generations, where each generation is a new family unit (father/mother, grandfather/grandmother, etc.) there would be a total of 2^n ancestors, where n is the number of generations. The current individual is generation 0, your parents are generation 1, grandparents are generation 2, etc. By 10 generations, a person would have 2^{10} or 1,024 ancestors. The more generations a person reaches back, the less likely it is that they will share DNA with the hundreds or thousands of ancestors. Each of us carries the DNA of some of our distant relatives, but not all or most of them. We have eight great-grandparents, and each of their great-grandchildren did not inherit the same DNA.

A Hypothetical Case of Isolated Populations

We shall begin with a simple hypothetical. Imagine that the human population originated hundreds of thousands of years ago at distinct, separated, and isolated regions of the earth. For tens of thousands of years, imagine that there was no crossover from these separated regions as humans evolved. Because of environmental factors that were distinct to these regions, such as differences in nutrition and food opportunities, along with Darwinian natural selection, certain characteristics became identified with these regions. Those inhabiting regions with little sunlight, such as Northern Europe, had evolved with lighter skin. Human populations with darker skin had difficulties surviving in that region. In the regions close to the equator, humans with darker skin had a survival advantage because it afforded them protection from ultraviolet (UV) radiation. Melanin, which is abundant in darker skin, impedes the breakdown of folate (also known as vitamin B_9) from UV radiation. Folate is essential for human survival because it is required by the body to synthesize DNA and RNA and to metabolize amino acids. In contrast, in regions further from the equator with diminished sun, light skin provided an advantage over dark skin in absorbing sunlight and producing vitamin D. Also, with less melanin in the skin, UV light more easily breaks down folate for absorption (folic acid). Thus, vitamin D may be synthesized during UV exposure and folate may be degraded.

We know today from genome sequencing that all human beings are about 99.9 percent identical in their genetic makeup. Whatever differences there are among humans throughout the world can be found in the other 0.1 percent of their genomes, which can reveal information about the causes of disease, ancestry, and familial relations. Children have about 50 percent of their mother's DNA, which means they share identical DNA sequences within the 0.1 percent. Since the human genome contains 3 billion base pairs that make up the nucleotides – or the rungs of the double helix – then 3 million base pairs (0.1 percent) would account for the differences seen across human populations.

Based on our hypothetical example of human populations in different regions evolving in isolation for tens of thousands of years, people from each region would possibly develop a set of markers in their genome that connected a person to that unique region. If people left one region and their remains wound up in another, the DNA from those remains could trace them to their region of origin because of the unique genetic markers (i.e., mutations specific to that region). Alternatively, if we found the remains of any random individual from a specific region, we could identify the genetic identifiers of that region from their genome (Figure 3.1).

In our hypothetical example of human origins, early human populations were split into distinct and isolated places. Imagine two groups totally cut-off from one another, where all migratory exchange was impossible. They would diverge genetically. If populations were so segregated in the early evolution of humans, there would be unique mutations. In those populations, determining ancestry would be a straightforward process.

For example, the mutation at the DNA sequence GCCGTAT (where the letters G, C, T, and A represent guanine, cytosine, thymine, and adenine, respectively, the bases that comprise DNA) in the genome of people in population A shows a G in the fourth nucleotide, whereas the sequence in people in population B has an A at the fourth nucleotide (Figure 3.1). But humans are a migratory species and thus isolated populations did not persist. If populations 1 and 2 share genetic markers (Figure 3.2), we might consider their frequency in each population. Thus, if the sequence GCCATAT appears in both populations, as shown in Figure 3.2, say 80 percent in population A and 15 percent in population B, then a person having this sequence would have a higher probability of having ancestry from population A than from population B. In general, a person is more likely to have ancestry from a population in which the markers are in higher frequency.

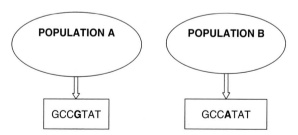

Figure 3.1 Distinct mutations in separated populations.

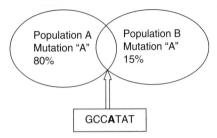

Figure 3.2   Shared mutations (genetic markers) in overlapping populations.

Our hypothetical about the origin of human populations (Figure 3.1) or what is called the polycentric (or multiregional) theory of human evolution, was not how humans were actually dispersed around the planet. Rather, the consensus theory is monocentric, namely that humans evolved from regions in and around East Africa for hundreds of millennia and eventually migrated to other regions of the earth (known as the "Out of Africa" hypothesis).

The Evolution of *Homo sapiens*

Based on fossilized skulls discovered in eastern and southern Africa, it is believed that humans evolved from the genus *Australopithecus*. Three species of the genus *Homo* are distinguished as having evolved: *Homo habilis*, *Homo erectus*, and *Homo sapiens*. The oldest of these, *Homo habilis*, is dated from 2–3 million years ago; *Homo erectus* is dated from 2 million to 300,000 years ago, appearing in eastern regions of Asia and surviving for nearly 2 million years. *Homo sapiens* (meaning "wisemen") are traced from 500,000 to 300,000 years ago. Modern humans are traced to the subspecies *Homo sapiens sapiens* (HSS) of the species *Homo sapiens*, dated around 100,000 years ago. A 2020 study in *Nature* reported on DNA data from four children, two buried about 8,000 years ago and two buried 3,000 years ago. Drawing from the data, the authors conclude their study with the finding that the presence of at least four modern human lineages that diversified about 250,000 to 200,000 BP (before present) and the DNA found in people living today offers support that this era was a critical period for human evolution in Africa. Depending on what the authors mean by "modern human lineages," modern humans can be traced between 100,000 and 250,000 years ago. Figure 3.3 illustrates the separation of the genus *Homo* from *Australopithecus afarensis*. Specimens providing evidence for this separation (including the famous "Lucy") were found in East Africa and carbon-dated to 3–4 million years ago.

Migrations began after humans spent tens of thousands of years in East Africa, building a rich human genetic diversity. It is generally accepted that anatomically modern humans advanced geographically from Africa and West Asia toward East Asia, Europe, America, and Australia.

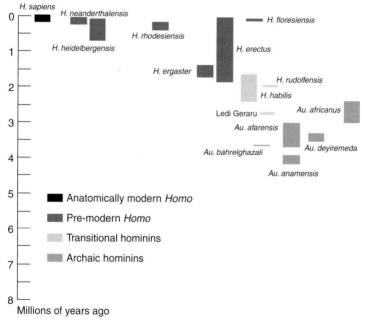

Figure 3.3 Descent of modern humans from Australopithecines. Source: Adapted from Wood, B. (2017). Evolution: origin(s) of modern humans. Evolutionary Biology 27: R746–R769.

Genetic Diversity in the Human Population

Around 55,000–60,000 years ago, HSS began migrating to Southwest Asia and Australia; 35,000–40,000 years ago to Central Asia and Europe, and 15,000–35,000 years ago to Northeast Asia and Australia. Each time HSS migrated, they carried some of that genetic diversity to the new areas they occupied, but they left behind a portion of the original genetic diversity, since those who migrated represented a subgroup of the original population. Once in these locations, the migrating populations developed new mutations, often unique to the region, and passed those mutations on to future generations. In these largely isolated populations, mutations were restricted through mating with closely related individuals. With smaller isolated populations, the mating will share common genes, resulting in less genetic diversity compared to the original population. Also, it was found that gene diversity in populations decreases steadily with increasing distance from East Africa. Following research led by Sarah Tishkoff at the University of Pennsylvania that compared diverse populations, it has been generally acknowledged that African populations have greater levels of genetic diversity than non-African populations.

In a classic paper published in 1972, population biologist Richard Lewontin, then at the University of Chicago, investigated human diversity within and between populations across the globe. Lewontin used protein electrophoresis (a method for separating molecules by weight) and immunological techniques to measure the genetic variation between individuals. Lewontin's evidence supported the conclusion that most of the genetic diversity (defined as the total amount of genetic variation) occurs within a population, while a much smaller amount of genetic diversity is found between populations. He wrote that people can observe variations among populations of, say, Asians, Africans, or Northern Europeans, where they identify phenotypes such as hair type, skin color, and facial bone structure, some of which are based on a single gene. But he noted that these variations are a small part of all the genetic variations in the human genome, which are not observable with the naked eye. Lewontin undertook his study three decades before the completion of the Human Genome Project that sequenced the human genome and two decades before the project was initiated. He had at his disposal 17 loci associated with blood groups. These loci had been characterized in a number of populations in which the alleles at the loci were found at different frequencies. For example, an allele CATTG at a particular locus on the genome appears at a certain frequency in the population. Another group of loci he examined are called serum proteins and red blood cell enzymes, which he studied by electrophoresis. The basic data he used were the frequencies of alternative alleles at various loci in two dozen population groups. The results that Lewontin discovered on genetic diversity within and between populations was in his terms "quite remarkable."

Lewontin found that the mean proportion of the total species diversity within populations is 85.4 percent, and less than 15 percent of all human genetic diversity is accounted for by differences between human groups. Genetic diversity is the variation in the amount of genetic information (total number of genetic characteristics) within and among individuals of a population. This tells us that if we wish to use genetics to distinguish two populations, we must avail ourselves of the relatively small number of markers in the genome. If humans differ by 0.1 percent of the genome, 15 percent of that (about 0.5 million nucleotides) can explain population differences.

These results have been validated by other scientists who have found similar or lower diversity between populations. In 2002, Rosenberg et al. used a different data set of genetic markers to Lewontin and reported in *Science* that within-population differences between individuals account for 93–95 percent of genetic variation; differences among groups constitute only 3–5 percent.

What we see when we look at individuals from different regions are some prominent features such as skin color and facial characteristics, and we take these to be the representation of genetic variability in the population. But this is a faulty generalization. Lewontin concluded in his paper:

> It is clear that our perception of relatively large differences between human races and subgroups, as compared to the variation within these subgroups, is indeed a biased perception and that, based on randomly chosen genetic differences, human races and populations are remarkably similar to one another, with the largest part by far of human variation being accounted for by the differences between individuals.

Even though Lewontin uses the terms "race" and "racial," reflecting common usage, he argues that racial classification has "virtually no genetic or taxonomic significance." His insights are prescient of the current consensus that race is a social construct and not a biological reality.

Thus, distinguishing population groups other than by genetics must utilize the most visible differences that often cannot distinguish between populations. Thus, Africans and Asians may be distinguished by eye, but that would not reveal most of their genetic similarities and differences.

Ancestry companies, however, have not totally discarded "race" as part of their analysis. When they claim to do so, it is often done incorrectly by affording "race" or its proxy a biological status. The public does not fully understand that there are not a set of genes for "race." At the University of Alberta, Kim TallBear, a scholar of racial politics in science and professor of native studies, examined the company DNAPrint Genomics (which went out of business in 2009). She wrote that the company relied on overly simplified notions that race has a biological basis. DNAPrint argued that race is found biologically in many visible differences between the populations of the world and that genetic differences have produced visible differences among people that we can objectively label as "race" differences. According to TallBear, DNAPrint Genomics focused on skin-color markers as a proxy for "race" markers. Dark skin color cannot locate an individual to the African continent since South Asian Indians, Southeast Asian Negritos, and Melanesians also have very dark skin color.

Lewontin's landmark study did not close the book on using genetics to identify differences among populations, even racial differences, with some scientists referring to "Lewontin's Fallacy." While they agreed with his conclusion that 85 percent of the genetic diversity was within a population, and barely 15 percent was between populations, they argued that the smaller percentage of diversity among populations contains evidence for distinguishing these populations. According to A. W. F. Edwards – a British statistician, geneticist, and evolutionary biologist – Lewontin fell prey to the "statistical fallacy" of analyzing data on the assumption that there was no information beyond that revealed by a locus-by-locus analysis. For Edwards, it is the correlation among the alleles in the distinct loci that can contain information about the ancestry of unique populations. The alleles and their frequencies have to be examined as a totality to distinguish unique populations.

The term "population structure" in population genetics relates to genetic diversity, and is defined by how those variants are distributed. Thus, allele frequency differences reveal population structure. For example, when the frequencies of a set of alleles cluster around one population at 0.86 and at another population at 0.20, they are said to reveal a different population structure (see Chapter 4).

The concept of allele frequency differences among population groups is the basis of DNA ancestry genealogy testing. Alleles that are common to all human populations and that appear at similar frequencies are not useful in distinguishing populations or their descendants. The method used to find this structure from small frequency differences in alleles can be traced to mathematician Harold Hotelling, who developed a mathematical procedure known as principal component analysis (PCA), a statistical technique used to analyze a large number of variables that are reduced to a small number of variables, called principal components, with a minimum loss of information. Currently, PCA is used to show patterns in genetic data and plays a special role in identifying the clustering of specific allele frequencies in human populations within geographical locations. Applying PCA to personal genetic data can be used to sort people into major continental population regions.

4 The Science behind DNA Ancestry Testing

Most human genetic diversity is found within populations rather than between populations. Scientists have reported that any two individuals within a particular population are as different genetically as any two people selected from any two populations in the world. Given this finding, how can science use a small percentage of genetic diversity between populations as markers of ancestral origins?

If each distinct biogeographical region consisted of individuals with a millennial lineage where members of that region had unique genetic markers, then the problem could be easily solved. Thus, if a single genetic marker at a locus on the human genome correctly and uniquely identified individuals within a relatively isolated region, then sequencing these markers would provide the genetic basis of ancestral genealogy. But there are no unique alleles in a population. What does exist are frequency differences of mutations (how often particular mutations occur) within populations (see Chapter 3, esp. Figure 3.2).

Allele Frequencies and Gene Variants

If you chose, say, six alleles and measured the frequency of their occurrence in two populations, the mean and the variance might be nearly the same. Thus, suppose in population A the six alleles have a mean frequency of 0.72 with a variance of 0.03 (80 percent of the cases fall between 0.69 and 0.75); this means that if you measured n people in population A for alleles 1–6, each allele with a frequency of occurrence (allele 1 = 0.70, allele 2 = 0.75, etc.), and you calculate the mean for the six alleles, you would get 0.72. Suppose population B has a mean frequency for those six alleles of 0.70 with a variance of 0.02. These values for a few alleles are very close. But if you sampled tens of thousands of alleles, the mean and variance for population A and B could vary significantly (i.e., a mean of 0.76 for A and 0.42 for B). Thus, the size of the population sampled and the number of alleles sampled are critical. For some researchers, even a modest number of loci would permit an accurate assignment of individuals to an ancestral population if enough polymorphic loci (several different sites on DNA for which different variants exist) are used. That is, there must be segments of DNA that are not identical across population groups or that vary in some unique ways. As an example, an allele known as the Duffy-null allele "FY*0," a gene variant that confers resistance to malaria, is almost entirely found in the genomes of those living in sub-Saharan Africa and is absent everywhere else.

Patents on Inferring Ancestry from DNA

The methods used for inferring ancestry lineage from the human genome vary among companies that have entered the commercial market. Patents on the components of the various methodologies were filed soon after the first companies were formed. These included patents on identifying common features in haplotypes (a group of alleles on a single chromosome that are so closely linked that they are not separated by the recombination of chromosomes and thus tend to be inherited from a single parent) using genetic markers to establish ancestral lineages. Patents also included methods of detecting genetic variation, discovering population structure from repeating genetic patterns, finding people descended from a common ancestor, and discovering ancestral origins for recently admixed individuals (people whose parents and their ancestors came from different regions of the world).

The smallest unit for a polymorphic (i.e., variable) site is a single nucleotide polymorphism (SNP). There are an estimated 15 million SNPs in the human genome. To be classified as a SNP, two or more versions of the sequence must each be present in at least 1 percent of the genome population. Only a relatively small number of SNPs have been classified as ancestry informative markers (AIMs), which are genetic loci showing large frequency differences between populations.

One of the earliest patent filings on a set of methods for inferring ancestry was dated November 18, 2004. The inventors are listed as Tony N. Frudakis and Mark D. Shriver. Frudakis is founder of DNAPrint Genomics and has worked in other medical and ancestry genetics companies; Shriver is a professor of anthropology at Penn State University. This patent application provides the scientific foundations and rationale for DNA ancestry testing in the context of what has been learned in population genetics over the past 25 years. They applied for patents on the method of inferring ancestry from AIMs, the probes for analyzing a person's DNA markers, and the forensic methods for analyzing crime-scene DNA for certain phenotypes, such as eye color or hair type.

I refer to the Frudakis–Shriver patent application as FSPA or "the patent application" in this chapter. Sixteen years after the filing, the patent was abandoned. A company had purchased their technology and was in the process of completing the application, but it was purchased from them before they could complete it. The company that acquired the patent application did not successfully complete the process.

The details of FSPA provide a unique view of how the innovators of ancestry DNA viewed the process. I focus on the patent application as a source of information because the section of the patent titled "Compositions and Methods for Inferring Ancestry" is the most transparent and comprehensive description I have seen, and it is unencumbered by layers of technical obfuscation. Moreover, they relate the DNA ancestry methods to developments in population genetics and were the first people to use autosomal DNA for ancestry. As part of the background information to their invention, FSPA states that the majority of the genetic variation among human individuals, which is around 80–90 percent, is "inter-individual," as we learned from the Lewontin study in Chapter 3. Only 10–20 percent of the variation is a result of population differences. Another way of saying this is that the substantial genetic variation occurs within a country or region, while the smaller variation occurs between countries or between regions.

Ancestry Markers

Frudakis and Shriver claimed that there are very few classical population markers, such as blood groups, serum proteins, or immunological markers that are population-specific or have significant frequency differences across geographical or ethnic populations. Yet they also maintained that the best opportunity to determine the biogeographical ancestry of an individual is through genetic markers in a person's DNA, based on the idea that unique environmental factors and nutrition can affect gene variants in a population, such as gene variants allowing for compatibility with particular diets high in alkaloids and tannins. They stated that their invention allows inference of ancestry, pigmentation, traits, drug responsiveness, and diseases susceptibility.

According to Frudakis and Shriver, the markers that reveal ancestry also reveal phenotypic properties of an individual (i.e., observable traits such as hair type, skin color, and disease resistance). They claimed that their invention could be used in forensics to narrow the population of suspects, in drug development to determine how individuals respond to medication, to ascertain an individual's likelihood of developing a genetic disease, and for genealogical purposes in determining the ancestry of an individual. It could be used to present an investigator with information on an individual's ancestry, including data on likely hair, skin, and eye color.

The patent application describes how the methods included were used to examine the crime-scene DNA of a murder-rapist in Louisiana and established that the individual was African American (85 percent sub-Saharan African, 15 percent Native American). They claim that the invention is predicated on the identification of AIMs, which serve as the genome components for ascertaining the population structure of individuals, which then allows an inference of traits. The alleles may be SNPs or deletion–insertion polymorphisms (DIPs), where a nucleotide is either deleted from or inserted into a sequence of DNA.

The patent application describes a method of analyzing a person's DNA to draw inferences about their ancestry. From a sample of a person's DNA, scientists locate the relevant SNPs. Then, with a panel of a minimum of 10 AIMs, scientists use a detection device that contains synthetic nucleotides

_____ Donor DNA strand
A G A G G T A T G G T T

T C T C C A T A C C A A   Hybridizing complementary probe

Figure 4.1   A hybridizing single-stranded DNA probe.

(segments of single-stranded DNA with the SNP) that will hybridize with the AIM of interest, indicating its existence with a fluorescent signal. The primers or probes in the detection device link to a polynucleotide that spans the AIM (Figure 4.1). The two strands will hybridize because they contain complementary base pairings (A–T, C–G). *Hybridization* refers to the association of single-stranded DNA through complementary base parings to form a double-stranded molecule. The sequences on the probe or donor DNA are labeled with an organic dye, and once hybridization occurs, can be read by various instruments. The probing methods are referred to by various terms, including *gene-chip*, *microarray*, or *bio-chip*. The chip is a platform for holding the synthetic DNA segments, and the probes are designed to detect haplotypes. The patent states that a specific SNP (e.g., 'A') is not informative in terms of the ancestry of the DNA section, but a haplotype (e.g., ACGA) starting at a specific location can be highly correlated with a specific ancestry (e.g., Northern European). Their detection method can be used on a panel of about 10 AIMS, a method that was subsequently expanded by scientists to thousands.

The patent application states that with a greater number of AIMs there will be more confidence in the ancestry inference. A similar argument holds for the number of loci used in forensic DNA analysis to determine the identity of DNA left at the crime scene: If you used 5 loci for the analysis, there would statistically be more than one person with those markers, but with 20 loci the exact match of the markers with a suspect would be unique. Similarly, the more AIMs one uses in a sample, the greater the probability of assigning a unique population.

One of the claims of the patent application is that the panel of AIMs can be used to estimate the biogeographical ancestry (BGA) of an individual, a collection of individuals at a population level, or at the subpopulation level (by ethnicities), as well as the micro-group level of families. The inventors cited evidence that a panel of 71 AIMs was identified from 800 candidate AIMs and that "methods were developed to examine these AIMs as a means to obtain accurate estimates of proportional ancestry." *Proportional ancestry* means the percentage of a person's DNA that can be traced to a specific region. The most useful alleles for determining ancestry including admixtures (individuals with DNA from parents of distantly related populations, also known as mixed ancestry) are those that exhibit large frequency differences among populations. The FSPA states that "a C or a G polymorphism at a particular place in the human genome, where the C is present in individuals mainly of European descent, and the G present in individuals mainly of Native American descent would have a high delta value, and therefore qualify as a good AIM."

In their statistical model, Shriver et al. introduced the Greek lower case letter δ (delta) to indicate the difference between the frequency of an allele at a specific locus on the chromosomes between two populations (px–py) where px is the frequency of one allele in population X and py is the frequency of the same allele in population Y. Greater δ levels indicate that two populations differed significantly in the frequency of an allele. They found differences for certain loci of 0.745 between Africans and Europeans and 0.524 between European Americans and Hispanic Americans.

The invention includes: (1) statistical methods for determining ancestral proportions in an individual's genome; (2) several hundred AIMs; and (3) software programs for determining ancestral proportions. The FSPA was the first patent application to describe the use of neutral autosomal SNPs distributed across the entire genome, not just a few chromosomes. Frudakis noted that they were not the first to measure "true" ancestry, but that they had used extensive validation to prove it, which had not been done before.

Other scientists describe the process of analysis whereby the donor DNA is broken down into segments along the chromosomes. The targeted sequences in the segments are isolated, but seldom are in sufficient concentrations for detection. The biological sequences of samples are amplified by PCR (polymerase chain reaction) methods, which create many copies of a DNA sequence so it can be analyzed. The DNA segments from the donor chromosomes are double-stranded. The probes for measuring the SNPs (if they exist at different loci) consist of synthetic DNA (the DNA sequences on the probe are synthesized) that are complementary to the SNP DNA strands at the loci, as described in the previous patent application.

Three types of DNA markers have been used to infer ancestry. Two of these are uniparental, which means they are passed down in a line of descent from either mother or father. These are mitochondrial DNA (mtDNA) or Y chromosome DNA. The mitochondria, which are organelles situated in the cytoplasm outside the nucleus of cells, consist of 37 genes and 16,569 base pairs, and encode 13 proteins and converts chemical energy from food into a form the cells can use. When the sperm and egg combine, only the female mitochondrial DNA gains residency in the embryo. Men inherit mitochondria from their mothers, but do not pass them on to their children. Mitochondrial genotyping has been used for inferring BGA as mtDNA is passed from mother to child. It is not mixed with the father's mtDNA, so it does not change very much from generation to generation. Mutations in mtDNA do not occur very often (~1 percent), so a person's mtDNA is likely identical to his or her direct maternal ancestor a dozen generations ago. This makes it suitable for inferring maternal ancestry.

The other uniparental marker is the Y chromosome, which is unique to males. An exact match of the Y chromosome of two men tells us that they share a paternal ancestor. Many of the early-formed DNA ancestry companies started off using mtDNA or the Y chromosome since they were relatively inexpensive to search. These two markers for ancestry have limitations, however. The mtDNA and Y chromosome tests only explore small segments of the genome (the matrilineal and patrilineal lineages respectively). Therefore, these tests only provide information for a very small proportion of a person's ancestors.

The third type of marker used for inferring ancestry is found in both males and females throughout 22 chromosomes (autosomes), excluding the sex chromosomes. This marker is a SNP, which we discussed previously. That there are SNPs in one population at higher or lower frequencies than in another has been noted by a number of researchers. These SNPs have been called "unique alleles," "population-specific alleles," and more recently "ancestry informative markers." The places on the human genome where a base can differ from one person to another is found at approximately 15 million SNPs, or about 0.5 percent of the genome.

Ancestry identification would be far simpler if certain populations had a mutation unique to their region. There are certain regions where some mutations are found at higher frequency than elsewhere, such as the hemoglobin allele (Hb*s) being most prevalent in sub-Saharan Africa. Those with two copies of the allele (called homozygous) have a high risk of being afflicted by sickle cell anemia, whereas those with a single copy of the allele (called heterozygous) do not contract the disease. Moreover, those with a single allele are resistant to infection by *Plasmodium falciparum*, which causes malaria. Using this mutation as an indicator of ancestry would not be specific enough for a region, since sickle cell disease has arisen throughout history among multiple populations, such as those in Mediterranean countries like Greece, Italy, and Turkey.

It has been noted that to distinguish populations, the ideal loci are those that have an allele located in one population and absent in another. However, such unique alleles have not been found and thus scientists have resorted to using clusters of alleles across many loci. According to Paul Kersbergen, two conditions are required to infer ancestry from a donor's DNA: (1) the availability of a suitable set of ancestry-sensitive markers (ASMs), also referred to as AIMs; and (2) a reference database with global frequencies of such ASMs. This will be illustrated in Chapter 7.

Validation of Ancestry Inference

Whatever method or methods are used to infer the ancestral lineage of people through their DNA, the ultimate validation of that method will be in its predictive accuracy. We can imagine such a test for validation. Suppose we select 100 people who have a complete genealogy of their family for five generations. That is, we select people who know the regions of the world their parents, grandparents, great-grandparents, great-great-grandparents, and great-great-great-grandparents were from. Then we apply a methodology to determine their ancestry from their DNA. The method would yield percentages of their ancestry from different regions. Finally, we compare the results of the method to the actual genealogies of the hundred subjects and determine how close the match is between the known genealogy and the DNA-generated genealogy.

While I am not aware that such tests have been run comparing predictive to actual genealogies and published in refereed journals, there is a lot of anec-dotal knowledge that people who get their DNA ancestry tests generally agree with the outcome, especially with regard to the major continental regions in which their recent ancestors are believed to have lived. Some ancestry companies have validated their own methods by selecting DNA profiles from their reference panels and using the reduced panel to predict the ancestry of the selected profile. One report noted that self-reported popula-tion affiliations correlated almost perfectly with the majority BGA of the population affiliation determined for a sample of 2,024 international samples. Pardo-Seco et al., in their paper "Evaluating the accuracy of AIM panels at quantifying genome ancestry," wrote that the ability of an AIM panel to measure ancestry can be evaluated empirically. Once we know the ancestry of an individual, we can compare that result with what is inferred from the AIM panel.

As previously stated, an AIM is an SNP that has substantially different fre-quencies among different populations. From an estimated 15 million SNPs in the genome, a subset is selected as a panel of AIMs. The ability of an AIM panel to measure ancestry is further complicated by the fact that the number of SNPs in an AIM panel can vary from a few dozen to a few hundred. As previously noted, the more SNPs, the greater the accuracy of the inference. Shriver et al. cited evidence that a panel of 100–160 AIMs provides reason-able outcomes. A number of studies cite a range of 42 to nearly 400 SNPS for determining a continental ancestry assignment.

Reference Panels

To summarize, SNPs occur at different frequencies within different popula-tions. The frequency differences of the SNPs reveal a unique population structure for a region of the world. Also, alleles at neighboring polymorph-isms may be associated and appear together. The science behind ancestry DNA   inferences   requires   reference   panels   for   each   population (biogeographical region) and a method of comparing the AIMs of an ancestry candidate with the reference panels of a region or population (e.g., Asia, Africa, Northern Europe). As previously noted, ideally the reference panel would be made up of the genomes of hundreds of individuals who are indigenous to the region and whose ancestors have lived in the region for hundreds of years (10 generations is about 250 years).

Scientists have shown that allele frequencies can reveal population structure with the aid of statistical methods such as principal component analysis. Once the reference panels are established for each region or population group, the ancestry detectives require a method of comparing the AIMs of the ancestry customer with those on the reference panels. Spencer Wells, author of *Deep Ancestry: Inside the Genographic Project*, writes that the reference panels would be composed of individuals living in a place all their lives, as their ancestors had done for centuries. They should be isolated from any migrations or immigrants into their region. By acquiring genetic data from indigenous populations around the world, we have a proxy for the genetic patterns as they were before new mass migration took place. The reference panels of companies are proprietary and thus are not available for scientists outside of the company to study. This has led some scientists to question the validity of the methods. Royal et al., writing in the *American Journal of Human Genetics*, noted that the unwillingness of companies to reveal their proprietary databases means their claims cannot be assessed, and so the reliability of the information they provide cannot be confirmed.

Allele Frequencies in Populations

The main criteria for the selection of a good AIM is the delta (δ) value, which is a statistical measure of the difference in minor allele frequency between various populations of human beings. The choice of markers that distinguish ancestry between populations is based on values of δ. To find SNP markers that are useful for ancestry testing, one has to genotype a number of samples from the major BGA groups of the world and find allele frequencies between at least two of the groups, calculate the δ values, and then rank them. The greater the δ values between two populations, the more the markers will be able to distinguish whether a person's ancestry is from those populations. For example, a polymorphism at a particular place in the human genome, where the C is present mainly in individuals of European descent and the G present mainly in individuals of Native American descent, would have a high δ value and therefore qualify as a good AIM.

Choosing AIMs that are used to compare a DNA sample with a reference panel is based on the differences in frequency they exhibit in different population groups. The frequency differences in the AIMs (markers) can also provide information about admixtures of different population groups. Table 4.1 shows a single marker and its frequencies in three regions of the world. The combination of numbers in "location" in Table 4.1 provide the gene's address from a reference point. In the example in Table 4.1, 8q13.2 is on the eighth chromosome, arm q, a distance 13.2 from the centromere, closer to the centromere than 8q16. The term "ter" refers to the terminus of the chromosome. Thus, 14qter refers to the tip of the long arm of chromosome 14.

The frequency 0.609 tells us that the marker appears much more commonly in the African compared to the European population. With respect to European versus Native American populations, the marker frequency difference is very low, which tells us that it is not a useful indicator to distinguish people's ancestry in these groups.

One of the complications of inferring ancestry from a person's DNA sample involves the stability of so-called indigenous populations. The reference panels are made up of individuals currently living in selected regions. They are chosen for the reference panel because they are reputedly indigenous, and thus it is believed that their recent ancestors lived in the same region. Once we get beyond a few recent generations, we cannot be sure that the markers represent people in the region. There may have been migrations from one region to another location or admixtures in the region from immigration – in other words, splitting or merging of populations.

Stages for Inferring Ancestry

The process of inferring ancestry from AIMs involves a number of stages between submitting a DNA sample and the final results being submitted to the customer. Figure 4.2 illustrates the stages of analysis. Not all ancestry companies used the designated AIMs sequencing for inferring consumer

Marker	Location	Mb	African/ European	African/ Native American	European/ Native American
CRH (NS)	8q13.2	73.2	0.609	0.655	0.046

Note: The first number (8) signifies the chromosome number (of the 22 autosomal pairs of chromosomes; the sex chromosomes are designated X and Y). Each chromosome has a p and a q arm separated by a constriction called the centromere, which is the reference point. The number indicating the gene position (13.2 in Table 4.1) increases with the distance from the centromere and is based on the position of bands when the chromosome is stained with a particular dye to make them show up.

Table 4.1   SNP markers and population frequency

ancestry. When asked whether his company used AIM markers, a Family Tree DNA spokesperson said that most methods, including their own, use a random sample of markers, not AIMs specifically. They use approximately 245,000 unlinked SNPs in their ancestry origins platform. The SNPs on the company's platform are used to analyze a customer's DNA. Once the SNPs are identified in the customer's genome, they are compared to the frequency of the SNPs in reference populations across the globe.

AncestryDNA is among the most transparent companies in explaining the science, the statistical methods, and the assumptions behind its inferences of its customers' ancestry. What follows is a list of their scientific assertions and assumptions, and a summary of the methods they utilize based on their 2018 "Ethnicity estimate white paper." This white paper, issued by AncestryDNA, is designed to help customers understand how their DNA is read and analyzed for the company to be able to draw conclusions about their ancestry.

1. DNA is passed down from one generation to another in long stretches of contiguous DNA (sequences) within chromosomes.
2. Their units of analysis are SNPs, single-letter changes (A, G, C, or T) in the DNA sequence.
3. Two nearby SNPs were most likely inherited from the same person.

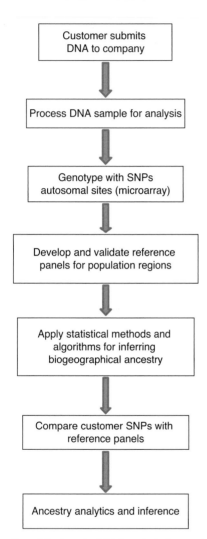

Figure 4.2 Ancestry DNA flow chart. Once customers submit their DNA (panel 1) the remaining steps are undertaken by the company.

4. They examine the SNPs in groups or clusters called haplotypes.

5. They assume that each haplotype comes from a population represented in one of their 43 reference panels (later expanded to 61).

6. They divide a customer's genome into 1,001 DNA sequences called "windows." Each window is 1–3 cM (centimorgans) in length, consisting of paternal and maternal DNA. A centimorgan is a unit of distance along a chromosome and represents about 1 million base pairs, but it is not a measure of distance in the ordinary sense of the term. A centimorgan is a 1 percent probability that two markers will cross-over on a single chromosome. Crossing over is a biological occurrence that takes place when the sperm and egg meet during meiosis and a new set of chromosomes is formed. Each male chromosome and its matching female chromosome are aligned and DNA segments are shuffled. A segment from the female chromosome breaks off and recombines with a segment of the male chromosome. This is why a child's DNA is somewhat different from his/her parents. When a 1 percent probability of cross-over is met, the distance between the markers is a centimorgan. Thus, if a marker on the X chromosome (identifying a gene sequence) and a marker on the Y chromosome, after meiosis and cross-over to a single chromosome, occurs more than 1 percent of the time, the distance between the markers is a centimorgan.

7. The haplotype within a "window" contains about 500 SNPs. It is assumed that the two parental haplotypes present in a "window" come from a single but not necessarily the same population, unless the ancestry of both parents derives from the same population.

8. The customer's haplotype within a "window" is compared to those in the reference panel to determine how likely it is that it came from any of the reference populations. For example, how likely do both sequences of DNA in a haplotype come from region A, or one from region A and the other from region B.

9. Each window gets a population assignment based on how well the SNPs match the highest frequencies found in a reference population. They compute the likelihood that the pair of haplotypes present in each window of a test sample comes from a population in their list of reference panels. Their computer programs consider all possible ethnicity assignments and accept the highest probability assignment. A customer

with the sequence Sweden/Sweden, Sweden/Sweden, Sweden/Sweden, France/Sweden, France/Sweden would be considered 20 percent France and 80 percent Sweden if the five "windows" are about the same size.

The section from the white paper addressing point 9 above explains how probability enters the analysis. When they analyze DNA data, they do not know the population it comes from ahead of time. They look at a pair of alleles (often called a genotype) at each position (SNP) in the DNA. One allele is inherited from the mother and the other from the father. Because the probability of a specific pair of alleles appearing at a certain position in the DNA varies for each of 61 regions, they use the information to estimate which region or stretch of DNA it most likely came from. If the pair AA at a particular position is more common in people from Spain, someone with AA at that location might have a higher chance of having Spanish ancestry. An AA at this position just makes it more likely the DNA comes from Spain. But according to AncestryDNA's assessment, people from Portugal, France, or even Korea might have an AA at this position as well. In Chapter 5 we look more closely at AIMs and their role in inferring ancestry from donor DNA.

5 Ancestry Informative Markers

If human populations in different geographical regions can be identified by DNA, while the vast amount of human DNA is identical, then, in the regions or places on the genome that are variable, there must be markers that reveal biogeographical regions of origin. DNA polymorphisms (letter changes in the nucleotides) are currently the choice markers because most human polymorphisms are characterized by alleles that are unevenly distributed among the world's distinct populations.

The concept of ancestry markers, often referred to as ancestry informative markers (AIMs), was developed, investigated, and validated in dozens of studies. According to a statement of the National Human Genome Institute on ancestry markers:

> Ancestry informative markers refers to locations in the genome that have varied sequences at that location and the relative abundance of those markers differs based on the continent from which individuals can trace their ancestry. So by using a series of these ancestry informative markers, sometimes 20 or 30 or more, and genotyping an individual you can determine from the frequency of those markers where their great, great, great, great ancestors may have come from. These are generally resolved to the three major continents: Africa, Asia, and Europe. Ancestry informative markers are used in epidemiological studies to see whether or not people have the same background, and sometimes they are used in forensic studies as well.

Forensic Applications of Genetic Markers

The idea of markers on the human genome that distinguish individuals was introduced in forensic science to help police to identify suspects. In 1983, Alec Jeffreys, a geneticist from the University of Leicester in the UK, created a method that used sequences of certain variable segments of DNA to establish the personal identity of an individual. Criminal investigators in the UK used Jeffreys' method to prosecute or exonerate suspects. A decade later, in 1994, Congress passed the DNA Identification Act, a section of the Violent Crime Control and Law Enforcement Act (P.L. 103-322), which gave the FBI the authority to collect DNA from convicted felons to establish their identity. The DNA markers were placed on a database called the Combined DNA Index System (CODIS). Four years later, all 50 states had authorized DNA databases. Police investigators could now upload the markers of DNA found at a crime scene to determine whether they matched a person in the CODIS database.

The markers used by the FBI and state criminal justice organizations are called short tandem repeats (STRs), which are hypervariable short repeated sequences – in other words, they vary a lot across human genomes. Initially, the FBI chose 13 loci across the genome, each locus being on a different chromosome and having two alleles. On January 1, 2017, the FBI expanded the number of loci to 20. These forensic markers have no known biological functions and STRs do not code for any proteins. Consider the following STRs from three different individuals at a locus on Chromosome 1 for each of two alleles.

Person 1   Allele 1: AGAC, AGAC, AGAC
                Allele 2: AGAC, AGAC, AGAC, AGAC, AGAC
                Number of STRs on both alleles 3/5
Person 2   Allele 1: AGAC, AGAC, AGAC, AGAC, AGAC, AGAC
                Allele 2: AGAC, AGAC, AGAC, AGAC, AGAC, AGAC, AGAC, AGAC
                Number of STRs on both alleles 6/8
Person 3   Allele 1: AGAC, AGAC, AGAC, AGAC
                Allele 2: AGAC, AGAC, AGAC
                Number of STRs on both alleles 4/3

The forensic scientific community has data on the frequency of paired STRs from the general population. As an example, person 1 has one allele with 3

STRs and a second allele with 5 STRs, designated 3/5. The second and third persons have 6/8 and 4/4 STRs at the same site. If we had the frequency of occurrence in the general population for each pair of STRs on 20 sites, we could calculate the probability that a single random individual would have this specific string of 20 STRs. In the above example, let us take five STR pairs for one of the individuals and *assume* they have frequencies of 0.02, 0.08, 0.05, 0.03, and 0.01. To get the probability that a random individual would have these five STRs we would simply multiply the probabilities:

$$0.02 \times 0.08 \times 0.05 \times 0.03 \times 0.01 = 240 \times 10^{-10} = 2.4 \times 10^{-8} \text{ or } 2.4 \text{ out}$$
$$\text{of } 100 \text{ million}$$

That result is only for 5 STR pairs. If we took 20 pairs of STR sites, the probability would be one in the many trillions. This would guarantee that if any individual's DNA matched the crime scene sample perfectly, there would only be one person in the world that could have that match since there are only 7.5 billion people on the planet.

The forensic STRs at the 20 sites on the genome are not supposed to reveal any phenotypic properties (physical characteristics) of an individual, although there are no guarantees that any of the STR sites are not linked to genes that do express traits. The markers sought by ancestry companies are quite different because they are chosen to reveal where people or their ancestors lived. Thus, markers for people who live or lived in Africa might be, but not necessarily, linked to levels of melanin expressed by a gene, indicating darker skin color. The criminal justice community was not only interested in DNA identification but also in DNA ancestry. Kersbergen et al., writing in *BMC Genetics*, noted that the ability to infer the genographic origin of a donor could add an extra dimension to criminal investigations.

From Forensic to Ancestry Markers

A number of papers published during the second half of the twentieth century by population geneticists presented models to estimate the admixtures of individuals from ancestral lines crossing different regions. One of these notable papers appeared in 1973 in *Nature*, and developed a mathematical model to estimate the admixture in African American populations from genetic markers.

The paper was based on a set of markers called "unique alleles." The paper by Chakraborty et al. provides a theoretical model for estimating the admixture of African Americans, who have both European and African ancestry. The analysis and model were based upon the presumed determination that there were 18 unique African alleles not found in Caucasians. These 18 unique alleles were the markers that establish African ancestry, where the level or percentage of admixture based on the distribution and frequency of the alleles determined the extent of admixture. The theory estimated the admixture of the African American population by the expected distribution of unique alleles among black Americans. They found that 18 variants can be used for studying the proportion of African ancestry in any population involving African admixture. The authors reported that these "unique alleles" are based on electrophoretic characteristics, where select proteins were measured by electrophoresis (passing the proteins through a gel in an electric field). They cautioned that electrophoretic identity does not imply identity by descent because the technique does not reveal genetic variation at the DNA level, but rather looks at the protein variations. Without the benefit of gene sequencers, these scientists had to use protein variants measured by electrophoresis. The authors recommended future studies on unique alleles to be done at the DNA level to support the validity of their results.

In 1973, James Neal, a geneticist at the University of Michigan, discovered genetic variants that appeared in different frequencies among South American Indians. He called these "inherited biochemical variants," and determined the frequency of these variants in six "relatively unacculturated" and "genetically pure" tribes of South America.

Mitochondrial DNA Markers

The first generation of DNA markers used for ancestry analysis were uniparental, as noted in Chapter 4 – namely, markers revealing only maternal or paternal ancestry. For maternal ancestry, stretches of mitochondrial DNA (mtDNA) were used. When an embryo is formed, it retains the entire maternal mitochondria, leaving aside the paternal mitochondria. Since female mtDNA are passed on to progeny without mixing with male mtDNA, the maternal line can be located across generations by markers on the mitochondrial genome, which doesn't change very quickly. There are hypervariable regions on the mitochondrial genome. These marker regions can be analyzed by sequencing the entire mtDNA – which contains about 16,500 bases, compared to 3 billion bases in the nuclear DNA – and compared to similar regions on reference panels. Except for spontaneous mutations in the transmission of the mitochondrial DNA, which are very rare, it will follow faithfully a maternal line across generations, since, unlike the chromosomes of offspring, the mtDNA is not a hybrid between the mother's and father's DNA. The marker regions in the mtDNA can also be analyzed by a microarray, which will be discussed in Chapter 9.

Individuals who share the identical sequence in a region of the mitochondria are said to belong to a haplogroup. They represent a maternal line of descent. Testing the DNA in a haplogroup is called a haplotype test. The markers are the hypervariable regions on the mtDNA. These can be used to infer maternal ancestry for both men and women, and are considered to be effective in tracing maternal ancestry for many generations. There are some limitations of mtDNA for informing anthropologists about the early stages of evolutionary history. Despite some limitations, even in the era of whole nuclear genome sequencing, the mitochondrial genome continues to be useful for the assessment of female-specific aspects of the history of human populations.

Y Chromosome Ancestral Markers

The second uniparental ancestry marker, as discussed in Chapter 4, is found in the Y chromosome, which is unique to males. In 1992, scientists discovered hypervariable STRs, similar to the type of markers used in forensic analysis, in the Y chromosome. This discovery had immediate applications in forensic science and paternity testing, and subsequently in genealogical DNA testing. The STRs are the short repeating sequences of bases (nucleotides), such as "TATT," that could be repeated 2, 3, 4, 5, etc. times in different people. The STR ancestry markers on the Y chromosome exhibit a paternal haplogroup – long segments of DNA inherited through the paternal line. Commercial testing kits can profile up to 17 Y chromosome STR markers, and a greater number of markers are expected in the future. As with mtDNA, these markers can be identified by a microarray or by sequencing the DNA on the Y chromosome, which has about 50 million base pairs. The Y chromosome can reveal information about the geographic region from where a person's paternal ancestors originate (i.e., biogeographical ancestry, or BGA).

While Y chromosomal and mtDNA polymorphisms were the first markers to be developed for both forensic and ancestry testing, they had their limitations. The most significant limitation of these sources of markers for ancestry determination compared to the use of autosomal DNA is that they only provide information on either the maternal or paternal ancestry, but not both. SNPs are more desirable as markers than STRs from a statistical standpoint because there is a larger number of them that can be analyzed simultaneously.

Autosomal SNP Ancestry Markers

The most widely used markers currently applied for multi-parental ancestry DNA testing are autosomal SNPs or microsatellites – segments of DNA with STRs across the chromosomes. These markers appear throughout the genome in all the chromosomes, with the exception of the sex chromosomes X and Y. In comparison to mtDNA and Y chromosome markers, autosomal markers provide much more comprehensive information on individual ancestry because cumulatively they represent a much greater proportion of the genome and thus its history (i.e., multiple biparentally inherited loci versus a single locus, as inherited through mtDNA or the Y chromosome). The main advantage of measuring autosomal over uniparental ancestry lies in the ability to measure admixtures within individuals contributed by all genetic ancestors rather than just some of them.

In 2010, none of the 24 ancestry testing companies reported using SNPs in their testing methods. By 2020, SNPs were used by the major DNA ancestry companies. Frudakis et al. believed that they developed the first SNP-based method for inferring ethnic origins of a DNA specimen in 2003. In their paper, published in the *Journal of Forensic Science*, the authors identified 211 SNPs associated with genes for phenotypic characteristics such as human pigmentation and xenobiotic metabolism because they believed the genes in which the SNPs occurred were subject to "unusual selective pressures over the course of evolution" and thus the SNPs would most likely show significant variations in different temperate zones. They chose 56 SNPs (mostly from pigmentation genes) and found that these frequencies were dramatically different between groups of unrelated individuals of Asian, African, and European descent, making them good for distinguishing among these regions in donor DNA.

Building on this early work on DNA markers, AncestryDNA wrote in one of its public information documents:

> AncestryDNA looks at about 700,000 markers in your DNA sample. Those markers are called SNPs (pronounced snips). Each SNP refers to a certain position in human DNA. And each SNP is made up of a pair of letters representing some combination of A, T, C, or G. Let's say that at SNP rs122 there are two possibilities: A and T. Because you get one letter (or allele) from each parent, you can have an AA, AT, or TT.

Some researchers cite far fewer SNPs to infer ancestry. Pardo-Seco et al., in their 2014 paper, state that panels with greater than 400 AIMs capture genome ancestry reasonably well, while those containing a few dozen AIMs show a large variability in ancestry estimates.

Nassir et al., in 2009, used fewer than 100 AIMs to distinguish a large number of population groups. They showed that a set of 93 SNP AIMs can distinguish a wide variety of diverse population groups in a sample that included the most populous groups in the United States, as well as groups from each populated continent with the exception of groups from Australia. Their methodology allowed them to use panels of AIMs for determining the origins of subjects from Europe, sub-Saharan Africa, the Americas, and East Asia. After they evaluated the performance of small panels of carefully selected genetic markers, it led them to conclude that the number of SNPs needed for ancestry inference can be successfully performed with less than 0.1 percent of the original number of SNPs while achieving close to 100 percent accuracy.

In 2010, Paschou et al. studied data sets of 1,043 individuals from 51 populations, including some from Africa, the Middle East, Europe, Central South Asia, East Asia, Oceania, and America, which contained 650,000 SNPs, obtained from the publicly available Human Genome Diversity Project. This project was publicly funded under the auspices of the Stanford University Human Genome Center. The genomic population data collected is available to all researchers and is referred to as the Human Genome

Diversity Cell Line Panel. The authors were able to reveal the genetic structure of different populations and maintained that fine-scale population differentiation was possible.

To reduce the complexity of ancestry analysis from hundreds of thousands of SNPs to a few hundred choice SNPs, scientists applied methods of marker reduction. In a 2009 paper by Halder et al., the goal was to create a single panel of AIMs that can be used to infer individual BGA and structure in different populations. By reducing the number of markers to a few hundred SNPs, the ancestry inference of a donor would be less expensive. But the trade-off of fewer markers is that the inference of ancestry would have less geographical precision that could distinguish different countries in Europe or in Africa. The authors in this analysis screened publicly available databases and identified candidate SNPs in ancestral populations. From a data set of 27,000 SNPs, they chose those with allele frequency >0.10 and each with $\delta > 0.4$ between any two groups (δ is the difference in allele frequencies between two populations). They ended up with 176 AIMs from four continental populations: Europeans, Indigenous Americans, East Asians, and West Africans. Their analysis demonstrated that they could infer the ancestry of a donor (if their self-reported ancestry was from one of these regions) to one of the four populations.

Soundararajan et al., in 2016, examined 21 published panels of ancestry informative SNPs to determine the level of uniformity. What they found was that only 46 (3 percent) of the unique 1,397 SNPs found in all the panels occur in three or more panels. They then formed a data set of 44 SNPs found in all the panels, involving 4,559 individuals from 73 populations. They concluded that their analysis of this data set provided clear differentiation of only five biogeographic regions – sub-Saharan Africa, Europe and Southwest Asia, South Asia, East Asia, and the Americas – and that this is an inadequate level of biogeographical resolution that was already exceeded by other panels. The authors believed the current panel of SNPs to be sufficient to discriminate among the five continents. They recommend a common data set of SNPs and an expansion of coverage to smaller regions of the world. There are problems in obtaining genomic data from certain countries. For example, China, India, certain indigenous groups, and some other countries will not allow DNA samples to leave their countries. In other cases, some populations will not allow their DNA and genotype data to be shared. Some scientific papers have discussed the validation of ancestry SNP panels and the methods used to evaluate them.

Pardo-Seco et al. noted that the ability of an AIM panel to measure ancestry can be evaluated empirically by examining its performance on a given set of DNA samples for which a given ancestry is already known. They consider the AIM panel efficient if the inferred ancestry is consistent with self-reported ancestry and/or with genealogical records. In other words, how well do the ancestry panels predict the ancestry of documented individuals?

Kidd et al., in 2014, sought to identify a small number of SNPs that would be useful for identifying the geographic ethnic origin of an unknown sample. His group found population sample sizes for each geographic region averaging 50 individuals, using a variety of sources to obtain the genomic information. They also used a set of 55 SNPs analyzed in 73 populations. The authors concluded that these conditions gave them the ability to infer the major geographic regions for the ancestry of an individual. These regions are Africa, Europe, and Southwest and South Central Asia. For greater resolution, the authors argue they need more SNPs and additional populations tested in the finer-grained regions.

There is consensus over the effectiveness of AIM reference panels to distinguish ancestry among the major continental regions of the world. Pardo-Seco et al. note in their paper evaluating the efficacy of AIM panels that with 10,000 SNPs selected at random from an individual, they can infer genome ancestry with negligible error for the three "HapMap" populations CEU (European), CAB (Asian), and TRI (African). HapMap, the result of an international project, is a database of common genetic variants (i.e., SNPs) in humans representing various regions of the world. Family Tree DNA reported using a random sample of SNPs from a source containing about 700,000.

Once the ancestry markers are chosen and can be used to analyze a sample of DNA, the next step is to compare the markers' allele frequencies with people indigenous to different regions. All DNA ancestry companies use reference populations of the regions for which they offer ancestry comparisons. The next chapter looks at the reference panels.

6 Ancestry DNA Population Reference Panels

In order to locate people's ancestry to a region of the world through their DNA, the markers on their DNA sample have to be compared to population reference panels for the regions that form part of the comparison group. These ancestry inference methods have served medical research, forensic science, and commercial popular genealogical interests. According to Santo et al., the reliability of any ancestry inference depends on the existence of reliable population reference databases. Many researchers and ancestry DNA companies utilize different sources for population data on different countries. For example, ALFRED is an allele frequency database supported by the Yale Center for Medical Informatics, which has genomic data from population samples across the globe. You can enter the name of a country or population group, such as Siberian Yupik (the sample was collected from unrelated Siberian Yupiks from northeastern Siberia, Russia) and it will provide information on the number of people (29) and/or chromosomes sampled (58).

Medical researchers are interested in discovering which population groups have higher frequencies of certain disease-bearing genetic mutations. They would also like to understand how different genomes respond to drugs – namely, how the nature and prevalence of pharmaco-genetically relevant polymorphisms can be estimated and whether genetic markers can predict an adverse drug effect.

The forensic community's interest in ancestry DNA reference panels relates to the role of DNA in solving crimes. When crime-scene DNA cannot be identified and matched on existing national DNA databases of felons, police sometimes resort to ancestry DNA panels to classify the DNA of population groups in order to narrow the range of suspects.

Anthropologists have used DNA reference panels to study migrations of peoples over thousands of years, while population geneticists have studied genetic diversity among human groups over time and across regions of the world.

Finally, ancestry DNA reference panels have been the mainstay of family DNA genealogical services, which offer customers links to their ancestral roots, including the biogeographical regions from which their grandparents and great-grandparents came. Each of these applications requires population samples that yield genomic information unique to the region. The samples must represent people and their ancestors who have lived in the region over generations.

Proprietary Reference Panels

Ideally, the population panels would consist of the ancestry DNA markers of a representative number of individuals indigenous to a region of the world for each of the comparative biogeographical sites. The samples obtained of indigenous people of a country, region, or continent are usually convenience samples, in contrast to a random sample of a population. A random sample would ensure that the genome data is not biased to a small subpopulation of the region. To qualify to have one's genomic information on a population reference panel, candidates must assert that they have lived in the region all their lives and did not migrate there.

Initially, scientists seeking to obtain international genome data selected five continents, since the most contrasting patterns of genetic diversity in humans occur at an inter-continental level. But as the human genome data sets expanded, finer-grained regions were chosen within Europe, Asia, and Africa. Ancestry companies followed by expanding the number of reference panels they used for individuating ancestry to smaller geographical areas, including countries or even parts of countries, such as areas populated by Siberian Yupiks. The inference of an individual's ancestry is only as good as the reference populations and the allele frequency data on the SNPs (single nucleotide polymorphisms) being used.

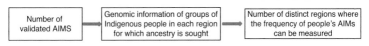

Figure 6.1   Requirements for inferring ancestry.

Establishing a model for inferring ancestry (Figure 6.1) of an individual requires a validated set of ancestry informative markers (AIMs), genomic information about people living in regions for which ancestry identification is sought (reference samples), and statistical methods for comparing the AIMs of a donor's DNA with the AIMs within diverse population groups. Here is how AncestryDNA defines a reference panel:

> A reference panel is made up of people with a long family history in one place or as part of one group. To make it into the AncestryDNA reference panel, these folks need two things. First, they need a paper trail that proves their family history. Then, they must have their ethnicity confirmed at the DNA level. It is not easy to make it into the panel! To tell your DNA story, we compare your DNA to the people in the reference panel and look for DNA you share. If some of your DNA is similar to the DNA of folks from Senegal, for example, we assign that part of your DNA to our Senegal region. And so on until we have looked at your whole DNA sample.

I have used the methods and analysis of AncestryDNA in this book for two reasons. First, it is transparent about its methods and assumptions. Second, it is among the two or three most successful companies.

In order to build more extensive and inclusive population reference panels, companies used genomic data from databases that were acquired by publicly and privately funded programs. These include the Human Genome Diversity Project (HGDP), ALFRED (the Allele Frequency Database, Yale), the 1000 Genome Project, sponsored by the National Human Genome Research Institute, and the International HapMap Project, an international collaboration of academic researchers, non-profit biomedical research organizations, and private companies. These early genome-mapping projects were focused largely on medical applications of the knowledge of genetic variation among populations.

The 1000 Genome Project was launched in January 2008. By November 2012, they had data on 1,092 people, in which they had identified 40 million variants. Currently, they have expanded their database to 2,500 people from 26 different populations. Some of the variants are more frequent in certain populations, which gives the genetic genealogists the grist for inferring ancestry from DNA. The company 23andMe developed 61 ancestry composition populations, which they described as genetically similar groups in different geographical regions. In their own words, they write:

> We select Ancestry Composition populations by studying the reference datasets, choosing candidate populations that seem to cluster together, and evaluating whether we can distinguish those groups in practice. Using this method, we refine the candidate reference populations until we arrive at a set that works well.

The company also uses the genomes and the personal information from customers to expand their reference panels. They noted that when a donor provides information that they have four grandparents all born in the same country that was not a colonial nation (e.g., United States, Canada, Australia), they include them on their reference panels. They describe how they use the reference panel:

> The Ancestry Composition algorithm calculates your ancestry by comparing your genome to the genomes of people whose ancestries we already know … Our reference data sets include genotypes from 14,437 people who were chosen to reflect populations that existed before transcontinental travel and migrations were common (at least 500 years ago).

Each company has its own set of proprietary population reference panels. There is not a standardized set of panels for all biogeographical populations. The SNP panels are developed from population genomics in a region – say, from Spain. When the allele frequencies are high enough compared to other regions, the SNPs are adopted to infer Spanish ancestry. There is a reciprocity between the SNP panels and the population panels for designated regions. Once the SNPs are derived from the genomics of the sampled population, then the algorithms find outliers in the population data (individuals whose allele frequencies depart significantly from the mean) and those outliers are removed. AncestryDNA writes that there are examples in which the person chosen in the population sample who identifies themselves as indigenous may be at odds with the genetic analysis of their descent, thus qualifying them as outliers.

In other words, even though a person lives in a particular region, they can be excluded from the reference population sample because their genetic profile diverges from the clustering of the principal components of the majority in the population. Imagine if we undertook a randomized political polling of a population and excluded all responses that were outliers (i.e., fell outside the standard deviation from the mean). We would not capture the diversity of responses. In this instance, the assumption driving the algorithms is that there are a core set of alleles that represent a population or region. Finding the diversity in the population is not the goal, whereas finding the core set of alleles is. The logic is that if the donor's SNPs match the core set, then they are more likely than not a descendent from that population.

In its 2018 white paper, AncestryDNA described its development of its population reference panels:

> Identifying the best candidates for the reference panel is key to providing the most accurate ethnicity estimate possible from a customer's DNA sample. Under perfect circumstances, we would construct our reference panel using DNA samples from people who lived hundreds of years ago. Unfortunately, it is not yet possible to reliably sample historical populations in this way. Instead, we must rely on DNA samples collected from people alive today and focus on those who can trace their ancestry to a single geographic location or population group. When asked to trace familial origins, most people can only reliably go back one to five generations, making it difficult to find individuals with knowledge about more distant ancestry. This is because as we go back in time, historical records become sparse, and the number of ancestors we have to follow doubles with each generation.

AncestryDNA also writes in its 2018 white paper on ethnicity:

> Because we use 43 global populations [since expanded to 61] in our reference panel, we divide the globe into 43 [also now 61] overlapping

geographic regions/groups. Each region represents a population with a unique genetic profile. Where possible, we use the known geographic locations of our samples to guide where the regional boundaries should be.

That white paper was amended in 2019, and AncestryDNA stated:

The AncestryDNA science team has developed a fast, sophisticated, and accurate method for estimating the historical origins of customers' DNA going back several hundred to over 1,000 years. Our newest approach improves upon our previous version in the number of possible regions that a customer might be assigned (from 43 to 61) as well as an increase in accuracy to both regions assigned and the percentage assigned to each region. We have added many new regions as well as made improvements to the composition of our reference panel, resulting in more accurate estimates overall.

Validation of Reference Panels

Some papers have discussed the validation of ancestry SNP panels and the methods used to evaluate them. In other words, how well do the ancestry panels predict the ancestry of genealogically documented individuals? As previously noted, those tests have not been reported in a form that lay readers can understand, although some studies correlate self-reported "race" with geographical regions.

While European ancestry genomic data is the most prevalent, scientists have begun expanding the data collection of regions within Asia. With 165 SNPs and 750 individuals from Asian countries, Lee et al. were able to validate reference panels that distinguish individuals from seven Asian populations and concluded that Southwest Asians have a genetic profile that is distinguishable from those of other Asian populations.

There is greater skepticism about the accuracy of inferring ancestry in smaller regions, as expressed by bioethicists Sankar (University of Pennsylvania) and Cho (Stanford), who wrote that the appearance of clustering is a function of how populations are sampled, of how the criteria for boundaries between clusters are set, and what level of resolution is used. They argued that in the same way the earth can be described by many different kinds of maps (e.g., topological or economic), so also can the genetic variation among populations be divided in numerous ways and be made to highlight any chosen similarity or difference.

As regions for ancestry analysis become smaller, migratory patterns and admixtures become more pronounced and make the genetic distinctiveness of the regions less significant for inferring ancestry – for example, in Singapore the AIMs are not as effective in distinguishing Chinese, Malay, and Indian subpopulations. But as the number of SNPs collected expands and the population samples grow, the mathematical confidence levels for improved ancestry inference also rise. Nelson et al. report on a commercially available panel of 500,000 SNPs in conjunction with DNA from 6,000 subjects. Their results show not only a high discrimination among African, East Asian, South Asian, European, and Mexican ancestry, but finer differentiation among Asian populations and African and African American groups.

AncestryDNA gives us as clear a picture of how they choose their population samples as any of the commercial DNA ancestry companies. Under perfect conditions they would construct their reference panel using DNA samples from people who lived hundreds of years ago. But these standards cannot be met. AncestryDNA maintains that knowing where someone's recent ancestors were born is a sufficient proxy for having the actual genealogical data of those ancestors. Thus, taking what people "know" about their recent ancestors is sufficient for AncestryDNA to identify a person's deeper ancestral roots rather than by other means. The company has drawn on population-based genomic data from the HGDP, the 1000 Genomes Project, and proprietary data, including the data from their customers when their family trees confirmed they had a long family history in a region or with a particular group. But AncestryDNA does not know what criteria or quality controls were used by HGDP and the 1000 Genomes Project for obtaining genomic data from individuals. They write in their white paper: "Although it was not possible to confirm family trees for HGDP and 1000 Genomes Project samples, these datasets were explicitly designed to sample a large set of distinct population groups representing a global picture of human genetic variation."

Table 6.1 shows the 61 regions for which AncestryDNA has population data, along with the number of samples (individuals sampled) in that population. The number of people sampled in each country or region does not correspond with the size of the population of that country or region. For example, in AncestryDNA's list (Table 6.1), there are 915 samples in China and 3,427 for

Region	Number of Samples	Region	Number of Samples
Aboriginal and Torres Strait Islander	14	Iran/Persia	577
Baltics	147	Ireland and Scotland	560
Basque	31	Italy	1,057
Benin and Togo	287	Japan	173
Burusho	23	Korea	197
Cameroon, Congo, and Southern Bantu People	535	Mali	413
Central Asia South	369	Malta	106
China	915	Melanesia	50
Indigenous Cuba	3,427	Middle East	467
Dai	70	Mongolia and Central Asia North	58
Eastern Bantu Peoples	91	Nigeria	522
Eastern Europe and Russia	1,777	Northern and Western India	413
England, Wales, and Northwestern Europe	1,461	Northern Africa	123.
Ethiopia and Eritrea	55	Northern Asia	43
European Jewish	450	Norway	402
Finland	408	Philippines	558
France	997	Polynesia	188
Germanic Europe	2,126	Portugal	627
Ghana	109	Samoa	73
Greece and the Balkans	354	Sardinia	38
Guam	57	Senegal	114
Indigenous Americas – Andean	62	Somalia	23

Region	Number of Samples	Region	Number of Samples
Indigenous Americas – Mexico	725	Southeast Asia	191
Indigenous Americas – North	2,290	Southern and Eastern Africa Hunter-Gatherers	38
Indigenous Americas – Southeast	2,966	Southern and Eastern India	405
Indigenous Americas – Yucatan	92	Spain	497
Indigenous Americas – Central	1,008	Sweden	414
Indigenous Americas – Colombia and Venezuela	2,615	Tonga	97
Indigenous Arctic	36	Turkey and the Caucasus	229
Indigenous Haiti and Dominican Republic	2,868	Vietnam	195
Indigenous Puerto Rico	4,803		
		Total	40,016

Source: www.ancestry.com/cs/dna-help/ethnicity/estimates; www
.ancestrycdn.com/dna/static/pdf/whitepapers/EV2019_white_paper_2.pdf
(accessed May18, 2020)

Table 6.1   The updated AncestryDNA ethnicity estimation reference panel contains 40,016 samples described to represent 61 overlapping global regions, each with a unique genetic profile

Indigenous Cuba. Also, we have no way of knowing whether the population samples are biased toward some allele frequencies over others.

To build their population reference panels, DNA ancestry companies may fund their own population samples, purchase private samples, or utilize

	ALFRED	AncestryDNA
Basque	100	31
Ghana	115	109
	1000 Genome	**AncestryDNA**
Japan	120	173
Nigeria	120	522
Finland	103	408
Puerto Rico	139	4,803

Table 6.2 Genomic samples by country: comparisons between ALFRED, 1000 Genome Project, and AncestryDNA

open-access databases. On examining ALFRED and the samples in the 1000 Genome Project, we can see that, in some cases, AncestryDNA does not use the entire set of samples from a particular country or exceeds the genomic samples of a country listed on the public databases. In Table 6.2, AncestryDNA uses one-third of the Basque database from ALFRED, but extends the data from the 1000 Genome Project for Japan, Finland, and Puerto Rico. Family Tree DNA cited four sources for its population reference panels: its consumer database; the HGDP; the International HapMap Project; and the Estonian Biocentre.

Once the AIMs are established and the population reference panels are set, companies need a technology to read the DNA sample of a customer. Genetic sequencing is far too expensive for each applicant. What makes the process technically and economically viable is the development of DNA arrays that can detect the SNPs without the need for full sequencing.

7 Comparing a Donor's DNA to Reference Panel Populations

As we noted previously, the science behind DNA ancestry requires that one compares the unique genetic markers on the consumer's DNA sample with the frequency of those markers in reference panels representing different regions of the world. When the field of DNA ancestry began, it was a scientific project that involved the search for biogeographical DNA. Scientists could use changes in the human genome to determine how ancient populations moved around the globe. The further populations moved across the globe and the more time elapsed (many thousands of years), the greater the number of mutations or genetic variants. Genetic ancestry began with a half-dozen distinct continental regions and with markers called hypervariable microsatellites, or short tandem repeats (STRs) of DNA, 2–6 base pairs in length. These microsatellites were considered ideal at the time because they had a high heterozygosity, which means two different alleles at a site. A site that has an AA is homozygous, whereas one that has AG is heterozygous. The more diverse the alleles, the greater the chance of distinguishing allele frequencies among populations. Initially, scientists used changes in the maternally inherited mitochondrial DNA (mtDNA) and the paternally inherited Y chromosome. That changed when autosomal markers were chosen for ancestry analysis.

DNA Variants

There are four types of DNA variations in the genome, described in Table 7.1 with their frequencies in the general human population: SNPs, STRs, short

Type of variation	Example of allele	Frequency of variation
Single nucleotide polymorphism (SNP)		
Reference DNA sequence	CCG TAG CAA GGA	1 per 1,000 bp (base pairs)
Substitution A for G	CCG TAA CAA GGA	1 per 10,000 bp
Single nucleotide insertion (T)	CCG TAG TCA AGG A	1 per 10,000 bp
Single nucleotide deletion (G)	CCG TAC AAG GA	
Short tandem repeats (STRs)		
Reference DNA sequence	CCG TAG CAA GGA	
Short tandem repeat	CCG **TAG TAG TAG TAG** CAA GGA	1 per 10,000 bp
Short indels		
Reference DNA sequence	CCG TAG CAA GGA	
Insertion of six nucleotide sequence	CCG TAG **ATA CCA** CAA GGA	1 per 10,000 bp
Deletion of six nucleotide sequence	CCG GGA	1 per 10,000 bp
Structural variants (SVs); copy number variants (CNVs)	Large deletions, duplications, inversions	4.8–9.5 percent of the human genome can be classified as CNVs

Type of variation	Example of allele	Frequency of variation
Inversion variant	AACCCCCG **GCCCCCAA**	Uncertain
Insertions/deletions (Indels)	TACCACG C/GGTC CTGTTAG/GGATTCTAG	1 per 10,000 bp
Short tandem repeats (STRs)	TACCACG C/GGTC GGACATGATGATGATGCCAT	1 per 10,000 bp

Source: Adapted from Ku et al. (2010) *J Human Genetics* 55:403–415.

Table 7.1   Types of DNA variations in the genome

indels, and structural/copy number variants. SNPs as previously defined are single nucleotide polymorphisms in which one letter changes in a DNA sequence. STRs are short tandem repeats. Short indels are insertions or deletions of six nucleotides. Structural copy number variants are large deletions, duplications, or inversions of DNA (Table 7.1).

A CNV is a type of structural variation that affects a substantial number of base pairs. Figures 7.1 and 7.2 show a large segment of DNA repeated on the duplicated chromosome. Between 5 and 10 percent of the human genome can be classified as having CNVs. These CNVs may include genes, where a person might have four copies of a gene rather than the usual two. These variants can be inherited or occur de novo (i.e., during the process of meiotic recombination CNVs can occur from errors of DNA replication or from a mutation).

In their 1997 paper titled "Ethnic-affiliation estimation by the use of population-specific DNA markers," Shriver et al. referred to the markers as *population-specific alleles*. These alleles vary in frequency across populations. They wrote that "The most useful unique alleles for forensics, admixture, or mapping studies are those which have the largest allele frequency differences among populations." A δ value equal to 1 would indicate no sharing between the populations of alleles, while a δ value of 0 would indicate total sharing of the allele. They found δ levels of 0.745 between Africans and Europeans and 0.524 between European Americans and Hispanic Americans. Both of these δ values are high enough to make inferences of ancestry between these populations. A high δ value between two populations for a small number of SNPs does not mean that the two populations vary significantly across the genome. Shriver et al. were one of the first groups to identify DNA variants in the form of SNPs that could be used to infer a person's ancestry.

Population-Specific Variants

In a 2011 article published in *Human Biology*, geneticist Noah Rosenberg posed a series of questions about genetic variability among human populations based on genetic markers. Among his questions were: (1) Do distinctive alleles exist for specific geographic regions that distinguish individuals in one group from those in other groups? (2) To what extent is it possible to

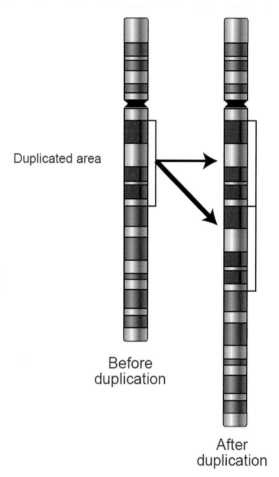

Duplicated area

Before
duplication

After
duplication

Figure 7.1   Copy number variants. Source: Wikipedia.

determine the genetic ancestry of an individual using the alleles in his or her genome? Rosenberg believes that most alleles are widely distributed around the world. About one-half of all the alleles represented in a panel are found in

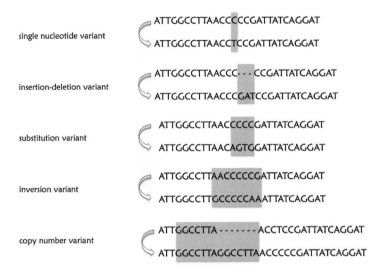

single nucleotide variant
ATTGGCCTTAACCCCCGATTATCAGGAT
ATTGGCCTTAACCTCCGATTATCAGGAT

insertion-deletion variant
ATTGGCCTTAACCC----CCGATTATCAGGAT
ATTGGCCTTAACCCCGATCCGATTATCAGGAT

substitution variant
ATTGGCCTTAACCCCCGATTATCAGGAT
ATTGGCCTTAACAGTGGATTATCAGGAT

inversion variant
ATTGGCCTTAACCCCCGATTATCAGGAT
ATTGGCCTTGCCCCCAAATTATCAGGAT

copy number variant
ATTGGCCTTA--------ACCTCCGATTATCAGGAT
ATTGGCCTTAGGCCTTAACCCCCGATTATCAGGAT

Figure 7.2 DNA variants. Single nucleotide variants are SNPs. Copy number variants are STRs. Source: *Kampourakis, Understanding Genes*, Figure 3.8.

Figure 7.3 Asymmetry between region and genome.

all seven geographical regions. Relatively few alleles are private to individual regions. Of those alleles that are private, more than half are found in Africa.

Rosenberg questioned whether distinct alleles in one geographical region could be used to distinguish individuals from other regions. He maintained that if the frequencies of alleles in a population are high, then these alleles could be used as diagnostic markers that could identify individuals as having ancestral roots in the population. If you knew the region a person was from, Rosenberg argues, it would still not be possible to predict the genotype of the individual (Figure 7.3). There is just too much variation within a biogeographical population. But if you had information about the alleles (SNPs) in a person's genome, it is possible to give an informed prediction of the region of their ancestry.

After several decades, beginning in 2000, commercial ancestry companies built up much larger reference databases, some with 45 distinct regions, including Melanesian, Siberian, Sudanese, and Filipino. By 2019, AncestryDNA cited 61 regions/populations to which they could infer ancestry of their customer's donor DNA. Companies also expanded the reference databases with donors' DNA that met certain criteria – for example, when donors reveal they have four grandparents, all born in the same country that was not a colonial nation (e.g., United States, Canada, Australia), their genome profile will be included in the reference panel of that country.

Suppose we had one SNP to work with and compared its frequency in five continental populations. Further, suppose one population had the SNP at a very high frequency, while it was absent in the other four regions. With some degree of confidence, we could infer that the donor's ancestry came, with high probability, from the one region that contained his or her SNP at a high frequency. Of course, that SNP might not have been inherited. Perhaps it came about from a mutational event that was unrelated to the ancestry of the individual. Also, we know that any single SNP may be found in any number of populations at different frequencies. We need to have many SNPs to work with to achieve some legitimacy for the use of allele frequencies in the human genome.

Population Frequencies of Variants and F_{ST} Statistics

Computational and population geneticists developed models and methods for investigating the allele frequencies of SNPs in a population and for predicting the likelihood that the SNPs of any individual donor corresponded to a particular biogeographical region. The most commonly used method is based on the statistical approach introduced by Sewall Wright in 1951, which he developed to describe the population structure of breeds of cattle, which was generalized to other species. From an intuitive standpoint, we are looking to ascertain the frequencies of a number of SNPs found on the donor's DNA for a series of biogeographical populations, taking one population at a time. A statistical quantity called F_{ST} was first suggested by Wright; it measures the amount of genetic variation within and among populations in terms of allele frequencies. It is calculated by finding the variance (the spread over the mean of numbers in a dataset) between gene frequencies for a set of n populations and then calculating the average gene frequency. If F_{ST} is small, the allele frequencies within each population are similar. If it is large, then there is greater differentiation of allele frequencies between populations.

Before we discuss the calculation of F_{ST} it will be useful to explain how allele frequencies are measured and how they relate to homozygosity (when an individual has two identical copies of an allele) and heterozygosity (when an individual has two different copies of an allele).

If N is the number of people genotyped, then the number of alleles is $2N$ (each person has two alleles at a locus). Suppose we calculate the frequency of allele A in the three populations (Table 7.2). If p = frequency of allele A, then for population 1, $p1A$ = the frequency of allele A in population 1; if q = frequency of a, then $q1a$ = frequency of allele a in population 1. Column 1 shows 125 people have two A alleles, which means that there are $2 \times 125 = 250$ A alleles in the population; column 2 shows 250 people have Aa alleles, which means there are 250 A alleles. Column 3 shows 125 people with aa alleles, therefore zero A alleles.

	Genotype			
	(1)	**(2)**	**(3)**	**N = population size**
	AA	Aa	aa	
Population 1	125	250	125	500
Population 2	50	30	20	100
Population 3	100	500	400	1,000

Note: "A" represents one version of a gene; "a" represents another version of the gene. "AA" or "aa" means the same version of the gene is on both chromosomes (homozygous). "Aa" means that each chromosome has a different version of the gene (heterozygous).

Table 7.2   A three-population example with allele frequencies

To calculate the frequency of allele A, we add the total number of A alleles divided by the total number of alleles:

$$V_p2 = å(p_i - p_{avg})2/n - 1$$

We do the same calculation for allele a:

$$\text{Allele a frequency in population 1} = \frac{0 + 250 + 2 \times 125}{2N}$$

$$= \frac{500}{1000} = 0.5 = q1a$$

$$p1A + q1a = 1$$

$$\text{For population 2}, p2A = \frac{2 \times 50 + 30 + 0}{200} = 0.65$$

$$\text{since } p2A + q2a = 1, \text{then } q2a = 0.35$$

$$\text{For population 3}, p3A = \frac{2 \times 100 + 500 + 0}{2000} = 0.35; q3a = 0.65$$

Now we have the frequency of each of the alleles in the three populations (Table 7.3). We shall see that the frequency of the alleles along the chromosomes is the basis of the methodology for inferring ancestry from a donor's DNA.

We can also determine the rate of homozygosity or heterozygosity for the three populations by calculating the number of homozygotes or heterozygotes in the population divided by the population number (Table 7.4).

The model developed by Jakobson et al. in 2013 seeks to show the connection between F_{ST} and allele frequencies. The population genetic statistic

Population	Frequency allele A	Frequency allele a
Population 1	0.5	0.5
Population 2	0.65	0.35
Population 3	0.35	0.65

Table 7.3   Allele frequency in three populations

	Homozygotes	Rate of homozygosity	Heterozygotes	Rate of heterozygosity
Population 1	250	0.5	250	0.5
Population 2	70	0.7	30	0.3
Population 3	500	0.5	500	0.5

Table 7.4   Rate of homozygosity and heterozygosity

called F_{ST} describes the allele frequency distribution in subdivided populations. A value of 0 indicates that two populations are genetically identical, while a value of 1 indicates that the populations are maximally different with respect to genetic diversity. The following are the results of the theoretical model that connects allele frequencies with ancestry through F_{ST} calculations.

For a group of populations and a single allele with frequency p_i in population i and p_{avg} (the mean allele frequency over those populations), F_{ST} is calculated as:

$$F_{ST} = Vp^2 / p_{avg}(1 - p_{avg}) \qquad (1)$$

where Vp = variance and p_{avg} = mean.

The variance is calculated as:

$$Vp^2 = \Sigma(p_i - p_{avg})^2 / n - 1 \qquad (2)$$

where n is the number of populations and the summation is from $i = 1$ to $i = n$.

From an intuitive perspective, if an allele has very similar frequencies within several biogeographical populations, its variance will be small and the F_{ST} value will be small. A small F_{ST} value means that the allele is not a good indicator for discriminating ancestry. Alternatively, if the F_{ST} value is large, some of the populations will have a high frequency of the allele and some will have a small frequency, which is what you want to infer ancestry. As previously noted, F_{ST} varies between the limits of 0 and 1.

Consider the case of two alleles in two populations as discussed by Jakobson et al. In this model we have two possible alleles at a locus on Chromosome 1. The alleles appear in populations at different frequencies, but since one or the other appears in every person, the sum of the frequencies must equal 1 (Table 7.5):

Allele 1   .... ACTTA**A**CG where the SNP is the sixth letter, A, in the DNA sequence.

Allele 2   .... ACTTA**G**CG where the SNP is the sixth letter, G, in the DNA sequence.

These SNPs will have different frequencies in populations A and B. For gene *MCM6* and SNP ID rs4954490 the following are the frequencies of allele 1 for Gambia and West Europe as given in Table 7.5.

Western Europe (CEU): allele 1 frequency $p = 0.798$; allele 2 frequency $q = 0.202$

Gambia (GWD): allele 1 frequency $p = 0.128$; allele 2 frequency $q = 0.872$

$$\text{Allele 1 average} = p1_{avg} = 0.463$$
$$\text{Allele 2 average} = q2_{avg} = 0.537$$

Now, applying formula (2):

$$\text{Variance}^2 = (0.798{-}0.463)^2 + (0.128{-}0.463)^2 \,/2{-}1$$

Population	Allele 1	Allele 2	Sum
A Western Europe	pA1 0.798	pA2 0.202	1
B Gambia	pB1 0.128	pB2 0.872	1
Absolute difference	$\delta_1 = 0.67$	$\delta_2 = -0.67$	–

Note: pA1 = frequency of allele 1 in Western Europe; pA2 = frequency of allele 2 in Western Europe; pB1 = frequency of allele 1 in Gambia; pB2 = frequency of allele 2 in Gambia. Note that the frequency of allele 1 plus the frequency of allele 2 equals 1. δ = difference in allele frequencies in two countries.

Table 7.5   Two alleles in two populations

$$\text{Variance}^2 = \frac{0.112 + 0.112}{2} = 0.112$$

Now, entering the variance in formula (1):

$$F_{ST} = 0.112/0.463(1 - 0.463)$$
$$F_{ST} = 0.112/0.248 = 0.45$$

The value of F_{ST} tells us that there is a fair amount of genetic diversity between Gambia and Western Europe based on the allele frequencies for gene *MCM6*.

We calculated the F_{ST} for one SNP and two populations (regions). F_{ST} can be calculated for many SNPs across pairs of populations or multiple populations to determine which markers can discriminate best among the populations. As the SNP frequency difference between two populations increases, the F_{ST} value also increases. If the variance is small, that means the range of frequencies will not extend much beyond the average. From formula (1) we can see that a small variance will mean F_{ST} is small and the SNP will not be very useful for using a DNA marker to distinguish an individual among populations.

There is a second method for calculating the F_{ST} for alleles in two populations based on the expected heterozygosity of the total population and the average expected heterozygosity within each population. In this calculation the F_{ST} statistic is determined by the following formula:

$$F_{ST} = (H_T - H_S)/H_T \qquad (3)$$

where: H_T is the expected heterozygosity of the total population, and H_S is the average expected heterozygosity from within each population. We calculate this for two populations and four allele frequencies from Table 7.1. Here are the key data points:

Frequency of allele 1 for Western Europe: $p1A = 0.798$
Frequency of allele 2 for Western Europe: $q2A = 0.202$
Frequency of allele 1 for Gambia: $p1B = 0.128$
Frequency of allele 2 for Gambia $q2B = 0.872$

Western Europe: $2pq = 2 \times 0.798 \times 0.202 = 0.322$
Gambia: $2pq = 2 \times 0.128 \times 0.872 = 0.223$

Average heterozygosity within populations of Western Europe and Gambia:

$$H_S = 0.322 + 0.223/2 = 0.273$$
$$H_T = 2 \times \text{average frequency of } p \times \text{average frequency of } q$$
$$\text{Average frequency of allele 1 } p \ (p_{avg}) = 0.463$$
$$\text{Average frequency of allele 2 } q \ (q_{avg}) = 0.537$$
$$H_T = 2 \times 0.463 \times 0.537 = 0.497$$
$$F_{ST} = 0.497 - 0.273/0.497 = 0.224/0.497 = 0.450$$

We can see that the two formulas:

$$F_{ST} = V_p^2/p_{avg} (1 - p_{avg}) \ V_p = \text{variance}$$

and

$$F_{ST} = (H_T - H_S)/H_T$$

yield the same value of 0.45 for F_{ST} .

Companies can calculate F_{ST} values for thousands of SNPs for two or more regions at a time and select the SNPs with high F_{ST} values for the primary ancestry markers. Once the ancestry markers (SNPs) are chosen, they can be detected in the donor DNA by microarrays (see Chapter 8). Based on the match of the test-taker's alleles with the reference populations, where they appear in high frequency, ancestry inferences can be made. The next chapter describes the technology and the science behind it used to read the donor's DNA markers so that they can be compared to the various population reference frames.

8 Probing Your DNA

When you purchase a DNA ancestry service, you are sent a kit containing instructions for submitting a DNA sample. Most companies provide a plastic tube, which they ask the test-taker to fill with saliva or cheek swabs, seal, and return. When the company receives your DNA test sample, it is processed for analysis. As noted previously, the vast amount of your genome does not distinguish you from other individuals. Therefore, your genome is broken down into segments of DNA that contain the alleles of interest, rather than it being fully sequenced. Here is how AncestryDNA describes the processing of the DNA samples it receives from customers:

> [T]o obtain a customer's ethnicity estimate, we divide the customer's genome into small windows. For each window we assign a single population to the DNA within that window inherited from each parent, one population for each parental haplotype. Each window gets a population assignment based on how well it matches genomes in the reference panel.

> We do not know the exact haplotype boundaries, which differ between people, but we can achieve a good approximation by dividing the genome into 1,001 small windows. Each window covers one section of a single chromosome and is small enough (e.g., 3–10 centimorgans) that both the maternal and paternal haplotype, the DNA from Mom and the DNA from Dad, in a given window are likely to each come from a single, though not necessarily the same, population.

Analysis of Customer DNA and Microarrays

The DNA segments, which include single nucleotide polymorphisms (SNPs), first have to be amplified to ensure that there is a sufficient amount of DNA for the accuracy of the analysis. Amplification is accomplished by polymerase chain reaction (PCR), a method that duplicates strands of DNA. Second, the DNA is denatured, which means double-stranded DNA is turned into single-stranded DNA in order that the probes can access and detect the SNPs. DNA sequencing could be used to read DNA segments to determine the polymorphisms, but this is expensive, time-consuming, and burdensome, particularly with thousands of customers seeking an ancestry DNA test.

The technologies that have made DNA analysis economically viable are microarray oligonucleotide probes. A microarray is a type of chip used to analyze large numbers (from hundreds to millions) of RNA, DNA, or protein molecules. On the surface of the chip are short, synthetic, single-stranded DNA or RNA sequences called probes, which are immobilized onto a solid substrate such as nylon membranes, glass slides, or silicon chips – called a "multiplex laboratory on a chip." When the chip, with its single-stranded oligonucleotides, is exposed to a solution of single-stranded DNA or RNA, those molecules in solution, which are complementary to those in the probe, will hybridize (link together). Complementary base pairing (hybridization) is the process whereby the molecular shape of guanine (G) always bonds with the molecular shape of cytosine (C) with hydrogen bonds. Similarly, adenine (A) always bonds with thymine (T).

Once the specific probe has found its complimentary single-stranded DNA, the researcher can activate a light signal from a dye molecule (fluorescein) attached to the sample molecules. Thus, if the DNA single-stranded probe were AGGAT, then it would hybridize to its complement TCCTA. If the segment in solution were not TCCTA but rather ACCTA, then it would not hybridize, indicating that the probe could not find its complement. If the probe was looking for a common sequence in the sample and failed to find it, then there was probably a mutation at that site.

The earliest DNA array, traced back to 1973, provided a method of isolating DNA segments containing a specific gene. Since the 1980s, fiber optics in conjunction with light-absorbing dyes have been developed as chemical sensors in analytical chemistry to detect specific chemicals, pH, or chemical concentrations.

The idea behind detecting biological chemicals with other biological chemicals immobilized on a solid substrate was reported in 1983 when antibodies were coated on glass slips. When cells with antigen receptors were exposed to the array of antibodies, the cell antigen receptors would bind to the matched antibody. The cell probe consisted of a 1.1 cm^2 plate with 100 "holes" into which antibodies were deposited. Once exposed to the array, only the cells with the complimentary antigens would remain intact. Other cells would wash off.

This method was expanded from antigens (proteins) to DNA probes (or DNA microarrays), large sets of nucleic acids immobilized on solid substrates, and developed commercially by companies like Affymetrix, Applied Microarrays, and Illumina.

By 2000, a new approach to the development of DNA microarrays was developed by a group at Tufts University, headed by Professor David Walt. Their approach involved synthesizing DNA molecules and attaching them to small polystyrene beads. Those beads were then placed on the ends of fiber optic strands, which had a small well etched on the tips of the fibers to hold the beads. The beads contained different DNA sequences. Here is how the technology was described in its early stages in a patent filed by Walt:

> When using an optical fiber in an in vitro/in vivo sensor, one or more light absorbing dyes are located at the distal end. Typically, light from an appropriate source is used to illuminate the dyes through the fiber's proximal end. The light propagates along the length of the optical fiber, and a portion of this propagated light exits the distal end and is absorbed by the dyes ... Once the light has been absorbed by the dye, some light of varying wavelength and intensity returns, conveyed through the same fiber or collection fiber(s) to a detection system where it is observed and measured.

These early fiber optic detectors were used as detectors of chemical concentrations based on the principle that the total quantity of light absorbed correlates with the quantity of the sample.

Optical Fibers and DNA Probes

The development of DNA microarrays advanced with the development of optical fibers. The fiber can be attached to molecules, which respond differently to light. Small molecules can absorb and emit specific wavelengths of light. Barnard and Walt introduced the idea of a fiber optic chemical sensor in a letter to *Nature* in 1991. Five years later, Walt was coauthor of a *Nature* paper on a fiber optic DNA biosensor, which was the prototype of the microarrays used as the probe for ancestry DNA analysis. The authors described their invention as synthetic oligonucleotide hybridization probes that were covalently immobilized at one end of an optical fiber 200 μm in diameter. Each had a different immobilized probe on its tip. The fibers were bundled together to form a multiplex DNA sensor.

Up to 50,000 microfiber optic strands can be bundled together. In each fiber optic strand a narrow well is etched at its tip. Inserted in the nanowell are 2 μm (0.002 mm) plastic beads (microspheres). Different DNA sequences are attached to the individual beads, each deposited in the well of a separate fiber. Each microsphere has a bioreactive agent, which might be a protein (a molecule consisting of a chain of polypeptides with many [>50] amino acids), an oligopeptide (a linear chain of a few [2–20] amino acids), or a DNA molecule attached to the microsphere. Each bioactive agent, including nucleotides, is associated with a unique optical signature (fluorescein labeled) such that any microsphere with that bioactive agent will be identifiable by its unique signature.

Each bead (microsphere) can contain single-stranded DNA, which mirrors the DNA sequence surrounding and including the SNP at a particular locus

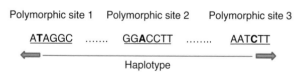

Polymorphic site 1   Polymorphic site 2   Polymorphic site 3

ATAGGC   ....   GGACCTT   ....   AATCTT

Haplotype

Figure 8.1   Polymorphic sites within a haplotype at a locus within a chromosome. The SNPs are T, A, and C, in bold.

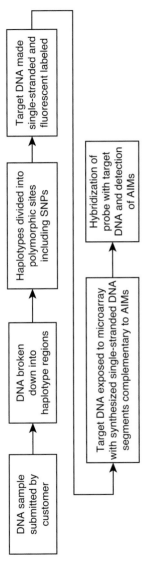

Figure 8.2   Flow chart for probing submitted DNA for AIMs in ancestry inference.

within a haplotype of the donor DNA. The fiber optic probe will detect the SNP if it is exposed to it.

Once the donor DNA is broken down into segments containing the SNPs, the probe can identify its complementary DNA segment when they are hybridized. Thousands of the fiber optic strands with synthesized nucleotides on their tips are placed on a solid substrate, which serves as the probe for SNPs on the target DNA. The array detector must contain the DNA segments of the ancestry informative markers (AIMs) that are used to analyze the donor's DNA.

The first microarray utilizing this method for DNA analysis was developed by the company Illumina in 2003 and had 1,536 DNA sequences. They began by selling the array to 23andMe. Prior to the fiber optic probe, companies were using sequencing and gene expression to identify variations in human genomes. In 2005, Illumina received a patent for its microarray, which it described as containing from two to many millions of different bioreactive agents on different beads.

The microarray is the detection system for determining which AIMs appear in the DNA sample submitted to the company. The sample DNA is broken down into haplotypes and then into AIMs within the haplotype. There are usually 3–7 polymorphic markers within a haplotype sequence (Figure 8.1).

The flow chart in Figure 8.2 illustrates the process from DNA submission to detection of AIMs in the sample.

9 Forensic Applications of Ancestry DNA Results

The criminal justice system began using DNA to solve crimes in the 1980s, after a geneticist from the University of Leicester in the UK developed a method for sequencing certain segments of chromosomal DNA. Those segments, called short tandem repeats (STRs), were expressed differently in different people, in contrast to the 99.9 percent of our DNA that is the same, and thus could be used to establish an indicator of personal identity (see Chapter 1).

Discovery of DNA Identification Method

When the geneticist Alec Jeffreys used his method to determine paternity, a police inspector in the UK asked Professor Jeffreys if he would help them determine whether their suspect was responsible for a double murder. Jeffrey's technique was able to show that the DNA that was left at two murder sites came from the same individual, but it did not match the DNA of their suspect. This is the first reported case of a suspect proven innocent by DNA analysis.

The technique that Jeffreys developed was subsequently refined and became the standard for DNA identification in the criminal justice system throughout the world. With two alleles per site, initially, a DNA identification was reduced to 13 pairs of numbers (e.g., 5/7, 6/8) representing the number of repeats for each allele at a locus, where there are 13 loci, one on each of 13 chromosomes. In 2016, the number of loci was expanded to 20 – that is, 40 numbers in 20 pairs.

The loci chosen by the FBI for forensic DNA identification were specifically devoid of phenotypic information. As a result, the 20 loci or 40 pairs of numbers merely identify a person whose DNA has a match, but tells us nothing about how a person looks. The loci were chosen because they are in sequences that do not code for proteins and presumably should not be able to provide any additional information about the person, such as their hair or eye color. But genetic advancements have correlated these non-coding markers with some phenotypic information that could inform criminal investigators.

The purpose of the FBI's Combined DNA Index System (CODIS) was to upload crime-scene DNA to determine whether there was a direct match with a person on the database. Alternatively, if the police had a suspect of a crime, comparing his/her DNA identifiers with the DNA at the crime scene could show that the person was at the crime scene. Without such a match on CODIS or otherwise, investigators could not use DNA to implicate a suspect.

When crime-scene DNA does not yield a direct match on CODIS or the state versions of it, the police, when permitted by state law or regulation, resort to familial searches. Police set the search parameters so that the profile of the DNA at the crime scene will pick up close matches on CODIS. Those matches are likely to pick up family members of the unknown suspect. Police can then use traditional investigative tools to determine whether any of the families of the familial matches links them to a suspect in the crime.

The Grim Sleeper

A serial killer, colloquially called the Grim Sleeper by the media, who murdered at least 10 women in South Los Angeles and possibly more over a period of several decades, left DNA at several of the crime scenes. It took the Los Angeles police 25 years to arrest a suspect. When the police ran the DNA sample on the state's felony database containing millions of profiles, they could not find a match. Then they ran a low stringency search (where most but not all the markers matched) where they could identify relatives such as parents, siblings, cousins, or nephews. The familial search led them to a recently convicted felon whose DNA proved him to be the son of the serial killer. The suspect picked out from the familial search was convicted of 10 counts of murder.

Familial searches have become widely used by police investigators when there are no prime suspects from DNA left at the crime scene. As noted by Krimsky and Simoncelli:

> familial searches use statistical methods to generate suspects in a crime where there is no prime suspect. Under the method forensic DNA has been transformed from a precise tool for identification into a blunt instrument for using DNA similarities to troll for family members of a person who happens to be in a DNA data bank – even while there are no independent grounds of suspicion of those family members.

If there are no matches or near-matches on CODIS, then crime-scene DNA will not reveal a possible suspect or provide conclusive evidence that a suspect was at the crime scene. The role of ancestry DNA has transformed this limitation. Without a match or near-match on CODIS, police have other ways of generating suspects in a criminal investigation. There are three ways police can use DNA analysis to find suspects. First, they can apply what is called phenotyping, which analyzes the crime-scene DNA for genes of physical characteristics (phenotypes), such as eye, hair, or skin color. A company called Parabon Nanolabs has used these techniques to provide police with leads. Second, they can submit the crime-scene DNA to an ancestry company to determine the most likely continental ancestry (Africa, Northern Europe, Asia) of a person. The police can operate under the assumption that people who are descendants from different populations exhibit different physical characteristics. That finding still leaves the police with great uncertainty over suspects since it does not significantly narrow the population of possible suspects who might have committed the crime. A third way is to locate family members by uploading the crime-scene DNA on a public database or to request access to the databases of ancestry companies.

The Golden State Killer

This third way that law enforcement has used DNA is best illustrated by the case of the Golden State Killer in California. Between 1974 and 1986, a number of heinous crimes were committed in various locations in California, including 12 murders, 50 rapes, and 100 burglaries. Police believed that one person could have been responsible for these brutal crimes based on forensic methods and the DNA found at the crime scenes. There were no matches when the crime-scene DNA was uploaded onto CODIS, so the so-called Golden State Killer became a cold case and remained so throughout the 1990s and into the new millennium.

Beginning in 2000, commercial DNA ancestry was beginning to take hold in the United States as a recreational activity for families. Police investigators tried something that was quite unorthodox and controversial. Using genealogical consultants, they submitted the crime-scene DNA to a DNA ancestry company under a pseudonym. Then they uploaded the data and the results of the ancestry analysis to an open-source genealogy website called GEDMatch, which was founded in 2010. The website takes all the uploaded DNA by individuals who were tested by different ancestry DNA companies and uses algorithms to find people who are related by similar stretches of DNA. Once a list of possible relatives of the unknown person of interest (the source of the DNA) is generated, professional genealogists begin to build family trees of the known individuals by utilizing traditional genealogical methods with the hope that the person who left the crime-scene DNA will fit into one of these family trees.

Genealogists working for the police constructed 20 family trees containing thousands of relatives to trace the killer's great-great-great-grandparents, who lived in the early 1800s. Investigators pored through census records, newspaper obituaries, gravesite locators, and commercial databases to eventually whittle the possible suspects down to one, Joseph James DeAngelo.

The search for the Golden State Killer by these techniques took years. A cold case expert was brought into the serial murder case by the district attorney of Contra Costa County, California. The consultant initially used the search site Ysearch.org, which led to a suspect whose DNA did not match the crime-scene DNA. Finally, the consultant turned to GEDMatch.

GEDMatch describes itself as an open-source genealogy website that invites those who have already paid for their ancestry DNA analysis to upload their data and profile onto the site. They claim to have over 1 million DNA profiles on their site as of 2018. Because the site is open to anyone who places personal DNA information on it, genealogist and criminal justice consultant Paul Holes recognized that it allowed law enforcement access to many profiles without a court order.

When Holes retrieved a rape kit (a set of materials, including vaginal swabs, blood samples, sperm residue, and hair samples obtained by a medical examiner from a victim of sexual assault) through the Ventura County District Attorney he found the DNA to be in excellent condition. Once the DNA sample was submitted to a DNA ancestry company for analysis, the results were uploaded to GEDMatch under a pseudonym. The site applied its algorithms and found distant relatives to the source of the unknown DNA. The method used for determining relatives of an individual is termed "identity by descent." Genetic genealogists determine the level of kinship between two individuals by how much shared DNA they have. This is done by examining SNPs in the genome of an individual and comparing it to others in the population sample. In this case it would be comparing the SNPs in the crime-scene DNA with others who have uploaded their DNA to GEDMatch. The genealogy algorithms screen the SNPs on the genomes of pairs of individuals. The algorithm finds regions where they share one allele at every SNP and sets the criterion that the matched segments must contain about 500 SNPs over a length of 5–7 cM. Table 9.1 shows the relationship between the length of the shared segments and kinship.

Identity by Descent

In "identity by descent" it is understood that two people with enough DNA similarity inherit DNA sequences from a common ancestor. The longer the stretches of DNA that two people have, the more closely related they are. As they are separated further in ancestry where there have been more recombination events, the shorter their shared (IBD) segments will be. The SNPs are inherited in clusters – that is, long sequences of DNA are inherited together. Nucleotides that are closer to one another on a chromosome are more likely to be inherited together, while nucleotides that are far apart are more likely to be separated by recombination. This is called linkage disequilibrium.

Relationship	IBD segment length (cM)	Matching chromosome (%)
Identical twins	3800	100
Parents	3400	50
Full siblings	2500	37.5–50
Grandparents	1700	25
Aunts and uncles	1700	25
Great-grandparents	850	12.5
First cousins	850	12.5
Second cousins	212.5	3.13
First cousin twice Removed	212.5	3.13

Note: IBD, inherited by descent.
Source: Verogen, Forensic Genetic Genealogy with GEDMatch. http://smithpla net.com/stuff/gedmatch.htm (accessed June 29, 2020).

Table 9.1 Centimorgan values as a measure of kinship

Segments of DNA are measured in centimorgans (cM) (see Chapter 4). The company Verogen provides a useful table that connects centimorgan values to kinship (Table 9.1). Verogen, Inc. was established in 2017 to develop, sell, and support forensic assays and software on Illumina sequencing platforms. According to company documents, Verogen was spun out from Illumina in August 2017 and was the only company solely focused on providing next-generation sequencing, a platform that enables the sequencing of thousands to millions of DNA molecules simultaneously, also known as high-throughput sequencing. Verogen also acquired GEDMatch and provides forensic and genealogical services that utilize the open-access database.

In the search for the Golden State Killer, genealogical investigators found a second cousin on the paternal side and a half-cousin once removed on the maternal side, genetically equivalent to a second cousin, who would have 3 percent of the crime-scene DNA (see Table 9.1). A cousin once removed is equivalent to a second cousin. Your mother's cousin (i.e., her mother's brother's son) is your cousin once removed (a difference of one generation). Your mother's cousin once removed is your cousin twice removed.

The genealogy expert used the distant relative matches to create a family tree that went back five generations to the early 1800s. Five generations means that the source of the crime-scene DNA had 32 ancestors. The original ancestors might have thousands of descendants. Investigators narrowed down the "huge" family trees they had created on Ancestry.com by consulting the census track, old newspaper clippings, gravesite locators, and Nexis Uni (previously known as LexisNexis Academic), which offers news, business, and legal sources. Out of some 25 family trees, the one including DeAngelo contained about 1,000 members.

Once DeAngelo was a leading suspect, police acquired DNA found on items that DeAngelo had discarded and on public surfaces he had recently touched, and they were able to determine that his DNA was a perfect match to the original crime-scene DNA. On June 29, 2020, DeAngelo pleaded guilty to 13 counts of first-degree murder, committed across the state of California in the 1970s and 1980s. He was sentenced to 11 consecutive life sentences.

When the media learned about the case of the Golden State Killer and how it was solved using the distant relatives of the suspect through GEDMatch, questions were raised about privacy and the fact that relatives, without their consent, were used to catch a criminal. GEDMatch became sensitive to the criticisms that people who had uploaded their ancestry DNA data were not informed that it could be used in police investigations. GEDMatch issued a disclaimer on its site.

> April 28, 2019. While the database was created for genealogical research, it is important that GEDMatch participants understand the possible uses of their DNA, including identification of relatives that have committed or were victims of crimes. If you are concerned about non-genealogical uses of your DNA, you should not upload your DNA to the database and/or you should remove DNA that has already been uploaded.

Solving the Golden State Killer cold case activated a wave of other efforts to solve cold case criminal investigations. Parabon Nanolabs was founded in 2008, based in Reston, Virginia. It specializes in nanopharmaceuticals, but also consults for law enforcement organizations on applying genetic genealogy to track down criminals. Parabon also used GEDMatch to track down family structures of unknown crime-scene DNA. The GEDMatch data can be accessed without a warrant by law enforcement, who, with consultants like Parabon, can find family relations of an individual who left their DNA at a crime scene, without those parties even knowing that they are helping police in the search. As of November 2018, Parabon reported working on 200 cases, of which 55 percent produced leads for criminal investigators. In May 2019, Parabon claimed it was solving cold cases at a rate of one per week. But if it sounds easy, not so according to Greytak et al., who reported in 2019 that during decades-long cold case investigations, hundreds of thousands of individuals can be investigated before a perpetrator is found.

Phenotyping

Beyond the trolling of open-access databases to find the family members of an unknown suspect through his/her DNA, DNA ancestry is applied in another way in criminal investigations. It is called phenotyping, or the prediction of physical appearance by DNA. Police use DNA ancestry analysis to determine whether a person is likely to be European, African, or Asian, affording them a broad-brush biometric identifier. By identifying an unknown suspect's biogeographical ancestry, police can reduce the number of suspects. It has been described how DNA can be used to determine iris color, gross aspects of morphology, skin and hair pigmentation, and facial features.

In 2011, a mother and daughter were murdered in their apartment. Four years later, without the crime being solved, police tried phenotyping. The police spent approximately $4,000 to have the sample analyzed by Parabon Nanolabs, and the DNA phenotyped. Parabon determined that the suspect was 92 percent West African and 8 percent European, with black or brown eyes and freckles. With this information the police produced a computer-generated photo-image of what they believed the suspect looked like. The police brought in a few suspects based on Parabon's computer-generated sketch. However, the suspects' DNA did not match the crime-scene DNA and no arrests were made.

Once considered a marginal method of criminal investigation and of limited validation, phenotyping has found its place in forensic technology. Some studies report that forensically validated DNA test systems are available for the categorization of eye, hair, and skin color, and the inference of continental biogeographic ancestry. The measure of test accuracy varies within the range of 0.74–0.99 for eye color, 0.64–0.94 for hair color, and 0.72–0.99 for skin color, and those figures depend on the predictive models. Nevertheless, it has been adopted by many criminal investigation units in the United States and Europe.

The next chapter will address the Fourth Amendment privacy aspects of engaging in genetic genealogy searches with online open-access genetic databases, and issues of personal identity in consumer ancestry tests.

10 Privacy, Personal Identity, and Legal Issues

When individuals sign on to a DNA ancestry test, they understand that the company will undertake an analysis of certain segments of their genome, called ancestry information markers (AIMs). These segments can, under proper analysis, reveal their genetic descent from certain regions of the world.

But as previously explained, there is a secondary market for ancestry DNA testing and that market consists of companies and research institutions that depend on large, diverse genomic data sets. The consumers' DNA obtained from their submitted saliva samples can be sold to these entities to be used for research and drug development. The submission might also include personal information of the donor, which, without a name, might also be included in the genomic data to entities in the secondary market. Many consumers are unaware that their genomic information is being sold and used for other purposes. Do they have the right of privacy beyond the information given with consent to the ancestry company? Should the donor's consent be required for the sharing of the information to other interested parties?

Ownership of One's Cells and Genetic Information

There are historical examples where people have given up genomic or cellular materials for one purpose and find that it was used, without their consent, for another purpose. In medicine, there is a long tradition of surgeons preserving cancer cells for research purposes. The most notable of these cases is that regarding HeLa Cells. Henrietta Lacks was born in Roanoke, Virginia in 1920. After giving birth to five children, she sought treatment for cervical cancer at Johns Hopkins Hospital in Baltimore, Maryland when she was 31 years old. The cells removed from a tumor biopsied during her treatment were evaluated for how well they proliferated and survived in culture. At the time, no consent was required for the appropriation of the cell line of a patient. Ms. Lack's cells proved to be one of the most significant cell lines in the history of medicine and became a standard for biomedical studies. Ms. Lacks died from her cancer in 1951. Her family was unaware of how her cells were being used and their importance in research. They received no compensation for the financial value of the cell line, known as the "HeLa immortal cell line."

In a similar case, John Moore of Seattle, Washington was treated at the UCLA Medical Center and diagnosed with hairy cell leukemia in 1976. He underwent surgery and had an enlarged spleen removed. Quite unexpectedly, and after a dire diagnosis for survival, Mr. Moore recovered from the surgery and returned to work as a land surveyor. His physicians cultured his spleen cells, which they discovered had a number of proteins known to stimulate the immune system. They took out a patent on the "Mo" cell line, which was awarded in 1984. Mr. Moore filed suit to gain commercial benefits from his property – his cells. The State of Washington Supreme Court ruled against Mr. Moore's claim that he had commercial rights over his body parts, but nevertheless ruled favorably on his claim that the physicians had an obligation to disclose their commercial interests in his cells. So in the state of Washington you don't own your cell lines when you undergo an operation, but if the surgeons have a financial interest in those lines they need to inform you before they acquire financial gain.

What happens to people who sign up for an ancestry DNA test? They willingly give up their DNA sample to the company for analysis. Some ancestry companies offer individuals the option of allowing their sample to be used for research beyond their own personal ancestry analysis. If they choose not to, the company states it will not sell or offer their genomic information to companies or research institutions. Some consumers may be satisfied to permit the transfer of their genomic data to other entities if it is anonymized – that is, their name or other identifying information does not appear. But consumers of ancestry DNA tests should know that by using methods to identify relatives of an unknown person's genomic data along with existing public databases and publicly available genealogical records, a genomic sample can be de- anonymized.

Suppose we were given an anonymous genome and wished to find the name of the person by using social media and public records. This was done by Erlich et al. in 2018. The investigators selected an anonymous genome from a publicly available database and uploaded the genotype to GEDmatch. They reported that after about one hour of work, they identified ancestral relatives from publicly available genealogical records. After a full day of work, they traced the identity of the anonymous genome. It has been concluded that with sufficient expertise, resources, and will, it is possible to reidentify individuals from anonymized family history data. Preemptive actions such as name removal and encoding will not protect one against privacy breaches.

There is no federal agency that has oversight of the privacy considerations in ancestry testing, although states have their own genetic privacy laws that may apply. One might assume that the donor of a DNA sample has complete ownership of it and can decide how it is used and to retract its use after a period of time. There are laws, such as GINA (the US Genetic Information Non-Discrimination Act) that are premised on the privacy of personal genetic information.

Ancestry Companies and the Privacy of Genetic Data

Companies like 23andMe claim they require consumer consent before they will sell your genomic information to a third party for research or commercial purposes. But they do retain the right to share your genomic information with third-party entities that provide services to their genealogical work.

For those consumers who believe they can delete their data from a database once they have received their ancestral information, there are two things to consider. First, companies provide updates on their ancestry analysis based on larger reference databases, which is an incentive for people to keep their genomic information on the company database. Second, it is not always easy to delete your genomic data from a database.

Legal scholar Emily Sklar, of Seton Hall University, wrote that consumers are under a false impression that they can delete their genetic data from a DNA ancestry company's database. In some cases, they can request that their account be deleted from the company website and that the company discard the physical samples. But their genetic information or personal identifiers will be retained for a period required by the Clinical Laboratory Improvement Amendments (CLIA) of 1988, according to the contract of services.

Consumer privacy has become far more important as ancestry databases expand and as the forensic uses of ancestry become mainstream in criminal justice. Anonymized samples may be re-identified using multiple databases. The federal laws that protect genetic privacy apply only to health data and not ancestry data. However, some states are expanding their laws to incorporate all types of genetic data. According to Sklar, while DNA is sequenced for both health and ancestry data, companies that do genealogy testing are not required to comply with FDA, HHS, or CLIA regulations of laboratories because of a loophole that exempts companies that do not produce health information.

A report of the Congressional Research Service stated that "The results of genetic ancestry tests do not fall under the definition of 'protected health information' under the Privacy Rule [Health Information Portability and Accountability Act], and testing companies do not fall under the definition of 'covered entities' under the Privacy Rule."

Third-party DNA searchers may be a game changer. It involves the privacy of individuals who have not consented to having their DNA used in criminal investigations or their genealogy exhumed by the police when they are not suspects. According to one report, one in four DNA ancestry companies claim that they disclose genome data to law enforcement agencies. Legal scholar Hilary Kody warned in 2019 that the courts have been left out of this hornets' nest of privacy infringements in both open-access and private DNA databases. She writes that with the vast amount of commercially analyzed DNA samples available through the millions of DNA ancestry kits sold, it is only a matter of time before any person in the United States can be identified through a familial search of a third-party DNA database.

At the current time, there are no rules or limitations for accessing the DNA ancestry profiles on open-access databanks. While ancestry companies typically do not open their databanks to the public, a number will allow police investigations without a court warrant. Ancestry DNA companies vary significantly in their public statements on the need to consider the privacy of third parties. The privacy issues of third-party DNA databanks, while amply discussed in law review articles, have not been challenged and resolved in the courts. The privacy of health-related genetic data has been established by law. Ancestry DNA can often reveal health information; however, it has been left out of the legislative, judicial, and regulatory arenas in the United States, with only a patchwork of oversight in Europe.

Advertising Claims

As direct-to-consumer DNA ancestry tests began to draw consumer interest in the mid-2000s, it was the marketplace exclusively that determined which companies succeeded. For the most part, state and federal governments have established no regulatory oversight of the industry. Lee et al. wrote in *Science* that direct to consumers ancestry tests are in an "unregulated no-man's land" where there is limited oversight and few guidelines regarding quality, validity, and, interpretation. Many unvalidated claims have been made in advertising. There were no standardized methods for inferring ancestry from samples of DNA, and thus the results that people have received are usually somewhat different.

Aggressive advertising determined the few leaders in the field of ancestry DNA testing. The boom in testing, according to some observers, may be a direct result of how much companies spend on advertising. Figure 10.1 shows the rapid growth of the sector beginning in 2013. In 2020, AncestryDNA reported selling its service to more than 18 million customers.

Initially, DNA ancestry testing was purely recreational. The results, in percentages of one's ancestral roots, became a conversation topic for people to share with friends and family. As the tests were refined and the reference panels tested for their internal coherence, the ancestry tests began to interest criminal investigators and lawyers engaged in paternity suits.

At the top of the list of social concerns are privacy and the reinscription of race into science. We shall begin discussing privacy and turn to race in the second part of the chapter. It starts when people submit their saliva or cheek swabs to an ancestry company. As previously noted, the genomic data you send to an

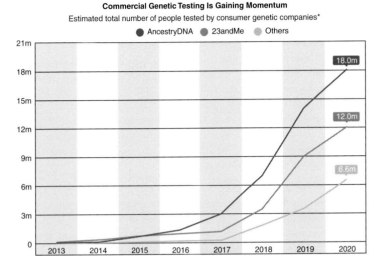

Commercial Genetic Testing Is Gaining Momentum
Estimated total number of people tested by consumer genetic companies*

● AncestryDNA   ● 23andMe   ● Others

18.0m

12.0m

6.6m

*Direct-to-consumer genetic testing uses DNA samples, such as saliva, to track a person's ancestry; find family members; disclose a limited array of possible health risks; or brief someone on their personal preferences, like a taste for cilantro or wine.
Sources: Leah Larkin, theDNAgeek.com.
@StatistaCharts

Figure 10.1   Growth in direct-to-consumer ancestry DNA testing.
theDNAgeek.com www.statista.com/chart/17023/commercial-genetic-testing/

ancestry company has value for pharmaceutical companies and research institutions. Your DNA, in combination with thousands of other samples, may reveal how patients respond to certain drugs. When ancestry companies sell your DNA and your personal profiles you might not know about it. And while it is likely anonymized, as previously argued, under the right circumstances it could be traced back to an individual.

Forensic Applications and Privacy

We saw in Chapter 9 how police investigators are using open-access DNA databanks to track down suspects. First or second cousins or cousins once removed who are linked genetically to DNA found at a crime scene may be placed under police surveillance and/or questioning to determine how they know or are related to the unknown source of the crime-scene DNA.

On open-access databases, there is a higher probability that a crime-scene sample will have fourth and fifth cousins genealogically than first and second cousins, since there are more of the former in an ancestry tree. But it is also the case that for some distant cousins (fourth or fifth) there will be no genetic matches. That is, genealogical cousins may not be genetic cousins, because they have not inherited the same DNA segments from their ancestors. For criminal investigators, it is only individuals who share three or more large genetic fragments with the crime-scene DNA who represent a strong lead in seeking suspects from a family tree.

In a family tree analysis, police investigators can find a number of distant cousins, but only a small number are likely to be genetic cousins, who are the ones you need to trace the identity of the owner of crime-scene DNA. First cousins, because one typically has a small number, are easy to identify. However, the probability that a random person has a distant cousin in a genetic database is low unless the database is very large. This explains why there was a degree of luck in tracking down the Golden State Killer through ancestry DNA.

In 2019, the chief biometric scientist at the FBI, Thomas F. Callaghan, wrote an editorial in the journal *Science* titled "Responsible genetic genealogy." He reported that the Department of Justice issued an interim policy on the use of forensic genetic genealogy. The policy does not allow investigators to use open-access genetic information, as if someone abandoned their drinking cup. The interim guidance limits investigative agencies to using only public databases or direct-to-consumer genetic genealogy services that forewarn users and the public that law enforcement may access their sites for investigative or unidentified human remains identification purposes. Without public confidence, Callaghan noted, law enforcement's use of forensic DNA analysis could be undermined.

Other social and ethical questions raised by ancestry DNA testing include the following:

1. Is there a need for consent of family members to use their DNA in a search of a suspect?

2. To what extent has ancestry DNA's population classification reinstated the concept of "race" in science?
3. Should consumers of ancestry testing have a right to determine whether their genome is sold to other companies for research or whether police investigators have access to ancestry DNA company databases?
4. Has ancestry DNA testing created confusion in the identification of human identity?
5. Has ancestry testing brought unexpected and unwanted information about family?
6. Is ancestry testing providing misleading information to clinicians about disease and ethnicity or race?

With respect to points 1 and 5, the ethical concerns include unexpected information that can upend families. An ancestry test might reveal misidentified paternity in a family. Children might be psychologically harmed by learning about false paternity at a young age. Such information can also impact inheritance. Genetic counselors Kirkpatrick and Rashkin believe that ancestry tests should include genetic counseling because of the overlap between ancestry and health information in the test. They predict that as the ancestry test marketplace continues to grow, and third-party vendors allow the general public to analyze their own genetic material, the role of the genetic counselor is likely to evolve significantly, which might include the interpretation of and the adaptation to the results. Others have argued that it has become increasingly difficult to separate genetic ancestry from information about health and that consumers of ancestry tests should be aware of the overlap and the privacy considerations.

Most people who send in their fee and a biological sample for an ancestry test are unaware that their results could be used by police to identify criminal suspects in cases where their DNA and the crime-scene DNA seem to be genealogically related. The question is: Do people have a right of privacy over their submitted DNA? There are two fundamental ideas regarding privacy that are recognized within US jurisprudence. The first is a person's "expectation of privacy." The second is the balance between privacy and the "state's compelling interest."

The courts have concluded that when a person enters a public phone booth, that the person has a reasonable expectation of privacy. As a consequence, police cannot wiretap the phone booth to eavesdrop on the conversation without a court warrant. We may also have an expectation of privacy in our home or in our automobile from searches and seizures by police. But if we are stopped by police for speeding, the courts have allowed them to search the vehicle on the grounds of "probable cause" of a violation in conjunction with the state's "compelling interest" in protecting public welfare.

If a person discards an object, the courts have ruled that they no longer exercise an expectation of privacy for an abandoned object. A paper cup thrown in the trash may contain the DNA of an individual. The police can, without a warrant, acquire the cup and analyze the DNA on it. When people submit their DNA sample to an ancestry genealogy company, it is unclear what their reasonable expectation of privacy is. If a company has a check-box in the registration document that asks people whether or not they mind having their DNA sample used for research without their name being revealed, we would have information about their expectation of privacy for the ancestry DNA sample. The fact that a person's DNA sample will be shared to other parties without a name does not negate their expectation of privacy for the biological sample. Even without a name, a biological sample could be de-anonymized, exposing an individual to a breach in the privacy of their personal genetic information.

Now let us suppose individuals upload their DNA profiles from an ancestry company on an open-access website. Have they given up their expectation of privacy of their DNA sample? Is that like exhibiting their DNA sample on their Facebook page? No, according to the editorial by FBI biometrician Thomas Callaghan. Currently there are no laws and policies to restrict the police from accessing DNA on an open-access website. And given that the police have found dangerous suspects from such searches, it is reasonable that the legislature and the courts might approve such searches as meeting a "compelling interest" of the state. At the current time, US law enforcement may search an open-access ancestry database to obtain a familial match without a warrant or regulatory oversight. This could lead police to distant relatives of the DNA source for whom there is no probable cause.

The FBI established CODIS (see Chapter 9) to match unidentified crime-scene DNA with that of convicted felons. They extend that to familial searches where similarities in the 20 short tandem repeats (STRs) imply a family connection. After police began using genetic ancestry databases, which, unlike CODIS, reveal phenotypic information from the genome, as one commentator put it, "the practical firewall between offender databases such as CODIS and commercial genetic databases is coming down."

Now police can combine their own DNA databases, CODIS, and commercial ancestry databases to develop and pursue suspects. According to Moreau, in the article "Crack down on genomic surveillance," "DNA databases in local police forces are proliferating, even in countries that have democratic governments and well-established legal protections for citizens' privacy." The author reports that the Office of the Chief Medical Examiner of New York City has a database of more than 82,000 genetic profiles, or about 1 percent of the city's population. For surveillance of an entire population you need profiles from 2–3 percent – other biological relationships can be inferred.

And what shall be said about the relatives of those who voluntarily place their DNA profiles in GEDmatch? Does a relative of a person who uploaded their DNA in a database have privacy interests in their shared genetic code? As one analyst concludes: "In essence, other relatives who share genetic information but never gave their consent for it to be exposed should not have their privacy rights ignored." The courts will eventually have to balance the benefits for solving crimes by accessing genetic ancestry databases against the privacy interests of all the individuals who are implicated. This includes those who have willingly uploaded their DNA profiles on a commercial database, as well as their relatives who have not uploaded their DNA.

Genetic ancestry testing has introduced new privacy concerns not anticipated by GINA, including the unintended effects of posting one's DNA profile on an open-access database and the concerns of family members who might become a suspect in an unsolved crime.

The motivation for people sending in their saliva samples for a DNA ancestry test may vary from seeking some answers to their genealogical roots, testing a family hypothesis of their descent, as evidence of minority status, or seeking to elevate their standing among their cultural peers. For some, the discovery of a bioregional connection centuries ago, far from where they currently live, can raise their self-esteem or give them a new perspective on their place in history. Much of how people respond to DNA ancestry tests depends, to a great extent, on the cultural values placed on their newly discovered identity. For some people, Native American heritage holds an enviable status and could confer benefits, even if a test reveals a mere 15 percent of it. According to sociologist Richard Tutton, senior lecturer at the University of York, the "rise of heritage" as a significant cultural and economic narrative in Britain has been linked with debates about national identity, the commodification of culture, economic decline, social change, and the resurgence of local identities.

Personal Identity through DNA Tests

During her 2018, senatorial re-election campaign in Massachusetts, candidate Senator Elizabeth Warren brought attention to her professed Native American heritage. The announcement brought a torrent of skepticism from her opponents and members of Native American tribes. Not long after the media blitz, Senator Warren reported her ancestry test results. Her ancestry DNA report was analyzed by Stanford University professor and DNA expert Carlos Bustamante and released to the media. He concluded that the results show Warren is mostly of European heritage but that the DNA indicates that she also has an unadmixed (pure) Native American ancestor from 6–10 generations ago. With that range, her Native American heritage could be between 1/64 and 1/1024.

It is generally acknowledged that each of the US Native American tribes determine the criteria for tribal membership. According to social scientist Adrea Korthase, there are 573 federally recognized Native American tribes in the United States, each with sovereign authority over its members. Some tribes require a certain blood quantum of citizens (fraction of their ancestors who are documented as full-blood Native Americans), while other tribes require no minimum blood quantum; some tribes require individuals to show a genealogical connection from a set of census rolls from the mid-1800s.

The issue of Warren's putative Native American identity was raised in a *Harvard Crimson* (the Harvard University student newspaper) article in

1998, when she was hired as a professor of law at Harvard. The context was that up until that time the Harvard Law School faculty had no minority women and she was cited in the *Harvard Crimson* as the first minority woman tenured at the law school based on her Native American heritage. According to the *Washington Examiner* (September 9, 2020) Warren had classified herself as Native American and was listed as a minority in a professional directory used by law schools for recruiting.

Warren's emphasis on her Native American heritage potentially could have elevated her approval among minority communities, who might not have focused on the percentage of her ancestry. But in her case the DNA as an indicator of her Native American heritage was disputed by tribal leaders. According to *New York Times* columnist Carl Zimmer, "Using a DNA test to lay claim to any connection to the Cherokee nation or any tribal nation, even vaguely is inappropriate and wrong." In terms of Native American identity, the general consensus among members of that community is that identity is not found in one's DNA, but is based upon one's social connections and lived experience with a tribal community.

Walajahi et al., in a 2019 article in Genetic Medicine, studied how ancestry DNA companies marketed their tests to Native Americans deploying scientifically inaccurate terms like "Native American DNA," or "Cherokee DNA." They write: "Native scholars and tribal communities have been openly critical of the use of genetic ancestry to define who is Native American and who is not, particularly as it reduces Native identity to a collection of biological determinants, superseding important cultural and historical considerations."

In parts of England there has been a resurgence of interest in the Vikings who lived in those regions during the Middle Ages. This interest has resulted in Viking festivals and associations. Scully et al. note that affiliation with the Viking culture is a source of "coveted identity." DNA ancestry tests offer people a connection to Viking DNA. The research team interviewed people in northern England about their interest in their Viking genealogy and their personal identity and how it changed with the ancestry DNA results. From their results they observed that personal identity was not reducible to a DNA test:

From our findings . . . it is difficult to sustain an argument that knowledge of DNA results leads to a reification of identity, or the geneticization of the social to the exclusion of all else . . . new DNA information is incorporated into narratives of identity in a more subtle way, it becomes a placeholder, to be woven into a broader narrative of selfhood in relation to the past, in dialogue with past personal or familial narratives, to be drawn upon at appropriate junctures, or to be retained until further information becomes available.

Nonetheless, DNA ancestry tests purchasers are intrigued at the prospect that their results might connect them through mitochondrial DNA (matrimonial lineage) or Y chromosome DNA (patrimonial lineage) to a historical figure who lived centuries ago, like Joan of Arc or Genghis Khan. Jobling et al. note that people are willing to embrace the mythology that their DNA is descended from a powerful and enduring historical figure who changed history. Companies offer DNA tests that claim to link customers to past celebrities, including Marie Antoinette of France, Tsar Nicholas II of Russia, and the outlaw Jesse James.

Genealogical searches of ancestry are much more serious for members of tribal communities. For many years, the United States and colonized countries like Australia had treated tribal communities as second-class citizens, if they were citizens at all. Social discrimination forced many to leave their tribes and to blend in with the mainstream, discarding their allegiance to their ancestral culture. As a new social awareness and acceptance of indigenous Australians and Native Americans began to grow, tribal descendants sought a reconnection to their cultural roots. In some cases, governments made restitution for past colonial abuses, offering "affirmative action" opportunities in land allocation or employment. Citizens of a federally recognized tribe may be eligible for benefits in healthcare, college scholarships, or financial rewards from tribal operations of gambling casinos. Proving one's ancestry to justify benefits and restitutions was required for some government programs, and some people believed ancestry DNA testing would provide the evidence they needed.

In 2002, litigation was initiated for monetary restitution from industries that benefited from slavery. Plaintiffs used DNA ancestry tests to demonstrate that they could trace their ancestry to African slaves in the United States. The challenges were unsuccessful.

In another legal case, Ralph Taylor, owner of the Orion Insurance Company, submitted relevant documentation and was certified as the owner of a minority business enterprise (MBE) by the Washington State Office of Minority and Women's Business Enterprises (OMWBE). With MBE status, Taylor applied for federally funded contracts. An official of the OMWBE claimed his MBE status did not qualify him for federal benefits and asked for additional documentation, whereupon his application was denied. Taylor sued state and federal agencies for denying his application that his business be considered a disadvantaged business enterprise (DBE), which under federal law established in the early 1980s provided certain benefits to small businesses owned and controlled by socially and economically disadvantaged individuals. The plaintiff alleged violations of federal and state laws, including anti-discrimination statutes and the US Administrative Procedures Act, a law requiring standardized procedures by federal agencies based on facts and evidence. Mr. Taylor presented the results of his ancestry DNA test, which he received in 2010. The results estimated that he was 90 percent European, 6 percent Indigenous American, and 4 percent sub-Saharan African. He acknowledged that he grew up thinking he was Caucasian, but in his late forties he embraced his black culture, represented in 4 percent of his DNA.

In addition to submitting his own DNA to qualify for the benefits, Mr. Johnson also submitted the results of his father's test, which estimated that he was 44 percent European, 44 percent sub-Saharan African, and 12 percent East Asian. The OMWBE questioned whether Mr. Taylor was a member of the black American or Native American groups and they also questioned whether he was socially or economically disadvantaged, a requirement to meet the conditions of acceptance in the DBE program. The OMWBE rejected his application in 2014 when they decided Mr. Taylor failed to meet their criteria. Its decision was upheld in the Ninth Circuit's Appeals Court.

Jewish identity must be established in Israel for marriage or for fulfilling the criterion of "The Law of Return" that allows Jews from other nations to take residence and gain citizenry in Israel. Many Jews seeking to emigrate to Israel are from war-torn countries where traditional genealogical documentation is not available. Rabbis have looked to ancestry DNA to adjudicate claims. It was reported in February 2019 in the Israeli newspaper *Haaretz* that the Chief Rabbinate of Israel had been requesting DNA tests to confirm Jewishness before issuing some marriage licenses.

As DNA ancestry is becoming more mainstream and more widely acknowledged in criminal justice, law, and medicine, it will be increasingly used by individuals who want to demonstrate that they can qualify for minority status. In the case of Native Americans, blood quantum as well as participating in tribal life have been the traditional measure of tribal identity, and DNA has not yet had – and may never achieve – a significant role.

Race and DNA Ancestry

The term "race" has largely been discredited as a scientific concept that helps to explain the origins or ontology of human populations. However, the methods that have been developed in genetic genealogy to classify human groups have reintroduced racial categories into the scientific lexicon. Do ancestry DNA tests reinforce people's racial identity and reintroduce it into science? Is the discovery of "racial" descent a desirable endpoint of DNA ancestry testing?

Over 50 years ago the anthropologist Ashley Montague dismissed "race" as a myth. However, racial categories are widely used in government policies, the census, and many other documents that do not distinguish science from the social adoption of "race." In the 2020 US federal census, people were asked to identify themselves within a list of traditional "racial" categories, including white, black, Chinese, Filipino, Asian, etc. The census document states that "the questionnaire reflects social definitions [of race] not biologically, anthropologically or genetically." The government wants census respondents to know that it is using the term "race" as it might be understood by the general public by relating it to country, region, or heritage of their descent.

There is little doubt that DNA ancestry tests have reactivated the term "race" in scientific discourse. People who undertake ancestry DNA testing are told which continental regions they are from and, more specifically, in which countries their ancestors lived. And yet the social concerns and science behind "race" are in some paradoxical way inter-related. Roth et al., in their article "Do genetic ancestry tests increase racial essentialism?" wrote: "Social scientists have long asserted that race is socially constructed, for instance, with classifications and their meaning changing over time and place, even if race refers in part to biological descent-based characteristics."

Sociologist Troy Duster warned us that scientific applications of "race" in medicine may contribute to its reification. DNA ancestry tests, which allegedly use science to draw inferences about genealogical descent, also reinforce essentialist categories of race in the public's mind. In *Race and the Genetics Revolution*, Duster wrote: "So, when it comes to molecular biologists asserting that 'race has no validity as a scientific concept,' there is an apparent contradiction with the practical applicability of research on allele frequencies in specific populations."

As we have seen, allele frequencies are the scientific basis for inferring ancestry from the DNA tests. It is possible to make sense of the apparent contradiction, according to Duster, "if we keep in mind the difference between using a taxonomic system with sharp discrete, definitively bound categories, and one which shows patterns (with some overlap), but which may prove to be empirically or practically true."

There is a new breed of population geneticists who have returned to racial distinctions through genomic data, although with many caveats. David Reich wrote in a *New York Times* op-ed: "I have deep sympathy for the concern that genetic discoveries could be misused to justify racism. But as a geneticist I also know that it is simply no longer possible to ignore average genetic differences among 'races.'" He also wrote that "'Race' is fundamentally a social category – not a biological one – as anthropologists have shown."

Some argue that genetic ancestry tests (GATs) are likely to reinforce genetic essentialist views of race. DNA ancestry results are in the form of categories that overlap with socially-defined racial groups. Therefore, they reinforce a view that these tests report race rather than simply biogeographical ancestry. Roth and Ivemark undertook a randomized controlled trial, with a sample of 802 native-born white Americans to investigate whether taking a genetic ancestry test increased the test-taker's views of genetic essentialism. They had one group that took the test and one group that did not. Both were surveyed on their views about race and genetic essentialism. What the investigators found was that in the treatment group (those who took the genetic ancestry test), changes in their genetic essentialist views were conditioned on their level of genetic knowledge – individuals with high genetic knowledge reduced their beliefs in genetic essentialism, and vice versa for those with low levels of genetic knowledge. The authors conclude that genetic ancestry testing tended to polarize test-takers by reinforcing their pre-existing beliefs about racial essentialism.

Roth and Ivemark also explored the role of genetic ancestry testing in racial identity formation. They refer to the "genetic determinism theory of identity formation" to describe "people who discover their genetic ancestral origins ... and view it as decisive proof of who they are ... including their racial identity." As an alternative they propose "genetic options theory," which states that "geneticization is not preordained, despite widespread speculation that GATs will advance the hegemonic role of science in shaping racial and ethnic identity." They argue that social forces play a key role in how people interpret the genetic science behind ancestry.

Based on their interviews, they found that their DNA ancestry testers viewed their ancestry as optional. But even as test consumers viewed their choices about personal identity from ancestry tests as not deterministic, the authors contend that the tests do reinforce racial categories in society. The authors concluded that the use of racial categories in ancestry testing may reinscribe race into science and in the public mind as an objective category for classifying people even as individuals pick and choose whether to heed their results.

Bryc et al., writing in the *American Journal of Human Genetics*, found that people choose the identity reflected in the majority of their DNA. Contrary to what one might expect under a social one-drop rule or "rule of hypodescent," which would mandate that individuals who knowingly carry African ancestry identify as African American, they found that individuals identify roughly with the majority of their genetic ancestry.

Genetic ancestry companies are aware of the sensitivity that many people have to reinscribing racial categories, yet they understand that some people even seek their racial identity. Horowitz et al. conclude that while companies carefully avoid language related to race, the way the services are advertised makes it easy for customers to believe the tests indicate racial identities. In 2015, Hochschild and Sen examined nearly 6,000 newspaper articles and concluded that "Americans are more likely to receive the message that genetic testing reifies or strengthens racial boundaries than to receive the message that DNA testing blurs or dissolves them."

In conclusion, DNA ancestry testing has brought genetic genealogy to millions of households. Consumers have largely embraced the results because they are putatively based on the science of DNA, albeit with the use of selected proprietary reference population data, statistical methods, and algorithms that are beyond the understanding of non-specialists. Among the values these tests have brought to test-takers is the confirmation of the family folk-wisdom of their lines of descent. For others, they have learned that there are parts of their lines of descent that were unexpected. Some test-takers believe that an actual document stating their ancestry can be beneficial to them in situations where ethnic and "racial" diversity or indigenous heritage are promoted in certain institutional and occupational settings. This includes the cases where Native American heritage can entitle people to certain legal benefits. Eliot and Brodwin, writing in the *British Medical Journal* explain that policy implications are at stake in ancestry testing because

> Determinations of ancestry or "blood" affect citizen rights throughout the world; the right of return of displaced people, membership in tribal bands ... and affirmative action eligibility ... Determining one's ancestry through genetic evidence would fundamentally transform these types of political identity ... family identity or caste.

In countries that experienced slavery for hundreds of years, with admixed populations, some consumers of DNA ancestry tests want to know whether the admixture of African and European populations left a mark of descent on them. Finally, we have discussed the ways that DNA ancestry tests reinscribe racial categories into science and the dangers behind such trends.

11 Discovering Unknown, Missing, or Mistaken Relatives

Much of recreational DNA ancestry offers consumers a long reach into the history of their descent by discovering which biogeographical population most closely matches their DNA profiles. The science and DNA analytics provide probability estimates that their DNA markers (ancestry informative markers, or AIMs) are most likely from a particular continent or even a specific country. But DNA ancestry tests have applications that go well beyond recreational genealogy. Even prior to the growth of this sector of direct-to-consumer testing, DNA was used to determine paternity and to establish identity in criminal investigations. An important and largely unintended application of ancestry DNA testing has been the uncovering of family secrets: "Why does my father look so different from his parents?" or "Why are my mother's skin tones so much darker than those of her parents?"

DNA and Paternity Claims

Paternity validation is currently a straightforward and dependable process. A child should have 50 percent of their father's and mother's DNA. While two children from the same father will each have 50 percent of their father's DNA, the segments shared may be different. Comparing the autosomal DNA of a child and a suspected genetic father will reveal whether he is the biological source of the child's DNA. The child also inherits his genetic father's Y chromosome. The two Y chromosomes should have considerable similarity.

But paternity testing was not always as reliable before DNA analysis. In the 1920s, blood typing was the only method available to ascertain whether the suspected biological father could be verified. Austrian immunologist Karl Landsteiner identified the A, B, and O blood groups in 1901. He discovered that there are red blood cells with type A antigen on their surface, which has antibodies against type B red cells. If type B blood is injected into persons with type A blood, the red cells in the injected blood will be destroyed by the antibodies in the recipient's blood. Similarly, type A red cells will be destroyed by anti-A antibodies in type B blood.

Blood typing was determined by those proteins on the surface of red blood cells. It was known that the blood type of the child was based on the blood types of the parents. Type O blood is universal – it can be injected without rejection into persons with types A, B, or O blood, because there are no A and B antigens in group O red blood cells, unless there is incompatibility with respect to some other blood group markers also present. Persons with type AB blood can receive type A, B, or O blood. Individuals may thus have type A, type B, type O, or type AB blood. The classification of human blood based on the inherited properties of red blood cells (erythrocytes) is determined by the presence or absence of the antigens A and B.

The blood type system could exclude certain claims of paternity. If the mother's blood type were A and the child's blood type were AB, then a male with blood type O could be excluded as the biological father. A child with blood group AB would have the genotype labeled $1^A 1^B$; he could then have inherited the allele 1^A from his mother with blood group A. Table 11.1 shows the possible alleles to the blood type. The gene

Genotype	Phenotype
$1^A 1^A$	A
$1^A 1^O$	A
$1^B 1^B$	B
$1^B 1^O$	B
$1^A 1^B$	AB
$1^O 1^O$	O

Table 11.1   Alleles and blood type

controlling the ABO blood group has three alleles: 1^A, 1^B, and 1^O. 1^A and 1^B are not dominant over one another, but both are dominant over 1^O. The blood typing chart in Table 11.2 illustrates the possibilities for the child's blood type, given the mother's and father's blood type.

If the mother has blood type B or AB and the child has blood type B, then the father could have blood type A, B, AB, or O, and thus no male would be excluded as the biological father (Table 11.2). Although blood group studies cannot be used to prove paternity, they can provide unequivocal evidence that a male is not the father of a particular child. Since the red cell antigens are inherited as dominant traits, a child cannot have a blood group antigen that is not present in one or both parents.

In the 1970s, blood typing was introduced based on proteins called human leukocyte antigens (HLA). These proteins were varied enough in the population to be used in paternity tests and were able to exclude falsely accused fathers. Since HLAs are inherited, where children have rare types, paternity could be established with high statistical certainty if the child's alleged father had the same rare type. When the HLA type was found more commonly in the population, the HLA test for paternity was inconclusive. Researchers who combined HLA tests with ABO blood typing and serological tests could increase the accuracy of paternity tests to as high as 90 percent. Currently, HLA typing is used to match patients and donors for bone marrow or cord blood transplants.

It wasn't until DNA was used as a method to determine paternity that the testing became highly dependable. The early method used *restriction fragment length polymorphisms*. These are fragments of DNA that vary in length across populations. The alleged father's and child's DNA in blood cells are broken down into segments by restriction enzymes. These enzymes cleave a DNA molecule into a short segment. The length of the fragments can be measured and visibly observed through electrophoresis. Matching the fragment lengths of DNA between the child and the putative father can establish paternity, since the child would inherit 50 percent of the fragment matches from each parent. When too many fragments do not match, the alleged father is excluded.

By the 1990s, polymerase chain reaction (PCR) had been developed as a method of duplicating small segments of DNA. The methods allowed

		Father's blood type			Child's blood type must be
		A	B	AB	O
Mother's blood type	A	A or O	A, B, AB, or O	A, B, or AB	A or O
	B	A, B, AB, or O	B or O	A, B, or AB	B or O
	AB	A, B, or AB	A, B, or AB	A, B, or AB	A or B
	O	A or O	B or O	A or B	O

		Child's blood type			Father's blood type must be
		A	B	AB	O
Mother's blood type	A	A, B, AB, or O	B or AB	B or AB	A, B, or O
	B	A or AB	A, B, AB, or O	A or AB	A, B, or O
	AB	A, B, AB, or O	A, B, AB, or O	A, B, or AB	
	O	A or AB	B or AB		A, B, or O

Table 11.2 Blood typing chart

paternity testing to be done with cheek swabs of DNA. The DNA segments are selected from specific loci of the genome and copied by PCR. The child's and the suspected father's DNA at the loci can be compared.

Typically, 16 fragments on the child's genome are analyzed and compared with those of a suspected father. If eight of them match closely, then it is a paternal match.

By 2000, SNP (single nucleotide polymorphism) arrays were developed with thousands of gene segments used as genetic markers. The DNA Diagnostic Center (DDC) uses an array with 800,000 SNPs, including AIMs, Y chromosome markers, mtDNA markers, and ancient DNA markers for establishing fourth- and fifth-generation cousins. DDC claims that with its newest sequencing it can determine the biological father of a fetus using blood samples from the mother (collecting embryonic cells) and a cheek swab from a possible father.

Hidden Family Secrets

The first line in a 2019 story published in *Harvard Magazine* stated: "On July 17, 2017, my world turned upside down when I discovered that the man who raised me was not my biological father." Stories abound about family secrets of hidden adoptions, illicit affairs and concealed pregnancies, unrevealed sperm donors, and surrogate-gestated children. Family secrets have a way of sustaining themselves for reasons of guilt, shame, or fear of losing kinship relationships. There was a time when it was common practice for parents to keep their adoption hidden from their child, but with the rise of open adoption laws and shifts in parental attitudes, not disclosing to a child that they were adopted is less pervasive in society. With the rise of ancestry DNA testing, people have been able to expose many hidden family secrets. In 2014, 23andMe estimated that 7,000 users of its service had discovered unexpected paternity or previously unknown siblings.

The *Harvard Magazine* story is written in the first person by Stuart Schreiber, a geneticist and co-founder of the Broad Institute in Cambridge, Massachusetts. Dr. Schreiber had experienced child abuse as a teenager inflicted by his well-educated, authoritarian father. He had unconditional love, however, from his mother. The fact that his other siblings were not also targets of his father's abuse had always puzzled Schreiber. The puzzle began to unravel after Schreiber's older brother shared his ancestry test results with him. With his genetics background, Schreiber was initially looking at disease risk factors in his brother's DNA profile, since their mother had died of Alzheimer's disease. Schreiber and his brother shared 25 percent of their DNA, which is not what is expected for full siblings. He also learned that their Y chromosomal DNA, which are near-identical in full siblings as inherited from their father, differed. That is when he realized that the person who raised them was only the biological father of one of the brothers. It was also when the selective abuse from his father began to make sense.

After conversations with his brother, Schreiber concluded: "I had lived with the sensation of being a family alien for 62 years, yet only at that moment did I realize it was true." The awareness that the father he grew up with was not his genetic father prompted Schreiber to find out who his biological father was. He spent months searching public and private DNA ancestry databases, constructing family trees across generations, and poring through newspapers. He solved the first part of the family mystery:

> In 1955, Joseph (Joe) was a handsome and charming young bachelor, recently returned from military service at the end of the Korean War . . . In my mind, Joe provided my abused mother with kindness and humanity when she was in great need – and I am the consequence.

Schreiber continued pursuing his genealogical roots and was able to assemble 150 "DNA-validated living family members." He constructed a family tree of more than 2,500 ancestors that goes back to his seventh great-grandmother, who he learned was the daughter of the chief of the Choctau Nation that settled the Mississippi Gulf Coast in the early 1700s. Through that inquiry he discovered the second secret of his family: His mother's father was not her biological father and that he had Native American DNA on his Chromosome 17.

From seeing his brother's DNA ancestry test, Dr. Schreiber finally exposed the secrets of his family heritage that his legal father was not his biological father and that his mother's legal father was not her biological father, thereby unearthing many hitherto unknown genetic relatives.

Family Ethnicity: Hidden Secrets

Alice Collins Plebuch took a DNA ancestry test at age 69 on a whim, expecting nothing unusual in its results. Her parents were Irish-American Catholics who raised six children. She hoped to learn something about her father where there were gaps in his ancestry. Jim Collins, her father, was believed to have been raised in an orphanage and was the son of Irish immigrants who could not afford to care for him.

When Plebuch received her ancestry test back, half her DNA indicated that she was descended from people in the British Isles, which was consistent with her Irish heritage. However, the second half of her genome matched populations in the Middle East and Eastern Europe, specifically Ashkenazi Jews. On first believing the tests results were wrong, she repeated the test and it confirmed what she had first learned, namely, there was a line of Ashkenazi Jews in her family. That's when Plebuch, who had skills in digital technology, began looking more closely into her family tree.

She purchased tests for members of her family to determine which side of the family – maternal or paternal – the Ashkenazi Jewish DNA came from. What typically comes into people's minds when they discover unexpected results in the DNA tests is that one person in the family was adopted or that someone was born through artificial insemination from an unknown donor.

Plebuch had her brothers tested and found that their X chromosome, inherited from their mother, had no Ashkenazi DNA, but the Y chromosome, inherited from their father, revealed Ashkenazi genes.

The ancestry DNA results from Plebuch's father's nephew began to solve the puzzle. Their father's nephew was not genetically related to Plebuch and thus to her father. That suggested that Joe Collins was not genetically related to his parents. Plebuch and her sister began reaching out to genome-sharing databases to see if they could find cousin matches on their father's side. With a considerable amount of traditional genealogical searching through birth records, the puzzle was eventually solved:

> Plebuch knew in her bones what had happened. This was no ancient family secret, buried by shame or forgotten by generations. This was

a mistake that no one had ever detected, a mistake that could only have been uncovered with DNA technology. Someone in the hospital back in 1913 had messed up. Somehow, a Jewish child had gone home with an Irish family and an Irish child had gone home with a Jewish family.

When a number of such mistakes were discovered, hospitals began using wrist bands on newborns to establish their identity at birth and to avoid baby swapping.

Secret Sperm

Children born by sperm donation, once they learn they have an unknown genetic father, frequently are curious about the donor. While sperm donations are common and donors are not difficult to find, many men will only donate under conditions of anonymity. US courts have protected those contracts involving both the donor and recipient of the sperm. In some countries those conceived with donor sperm have the right to acquire information about the donor. Most sperm donors, who believe they will remain anonymous, do not understand how the rise of direct-to-consumer ancestry DNA testing can circumvent their anonymity. Typically, the child born from a sperm donation who has an unwavering interest in determining the identity of the donor will first pursue traditional avenues, such as inquiries to the sperm bank or the physician who performed the IVF. If the protection of donor anonymity precludes acquiring the donor's identity, the next option is to request close genetic relatives from the ancestry DNA testing company and to post inquiries on open-access DNA databases to locate those relatives, such as genetic cousins, who may know the donor.

In one case, a journalist who was conceived through a sperm donation wrote that trying to get information from the sperm bank about his donor was incredibly frustrating and emotionally draining. He wrote that "I felt that they saw me as a problem they would have liked to go away." But he was not deterred and eventually was given some non-identifying information about his donor from his mother's clinic. It was through a DNA ancestry test that he finally identified his biological father.

One such case was highlighted on the popular ABC news documentary program *20/20* in a 2019 broadcast titled "Seed of doubt: Eve Wiley's story." Eve Wiley grew up in the small community of Center, Texas. Her legal father passed away when she was seven years old. One day Eve was looking at her mother's computer and read emails about sperm donors, with one seeking information about a man's daughter whose birth was the same day Eve was born. Eve confronted her mother, Margo Williams, and asked if she had something to tell her. That is when she learned that her legal father was not her biological father, saying in the *20/20* documentary:

> It's hard to imagine what it's like to come to the realization that something you thought you knew about yourself, namely, where you came from, is untrue. And that truth is about to take Eve from shock to heartbreak to fury.

Margo revealed that she pursued sperm donation when it was learned by her ob/gyn physician, Kim McMorries, that her husband's sperm by itself or mixed with sperm from the sperm bank could not conceive a baby. Margie tried again, choosing one sperm donor from 40 different offerings (known as Donor 106).

In adulthood Eve was able to learn the identity of her sperm donor (Steve Scholl) from the sperm bank. It seemed like the mystery of her biological father was over. Eve then turned to ancestry DNA with the hope of finding some siblings or half-siblings, since Scholl had donated sperm a number of times. By checking the box that asked if she wanted to find relatives, she learned from the company's database that her DNA matched to half-siblings. She delightfully reported that to Scholl, aka Donor 106.

Scholl then took his own ancestry DNA test. Eve began communicating with people she believed were her genetic relatives, matched through Scholl's DNA ancestry company. She discovered a genetic "cousin" who lived in East Texas, a place where neither Scholl nor his family lived, but where her mother went for her fertility treatments. Margo remembered that she was adamant about not having a sperm donor who lived locally. The people Eve found through social media on open-access genetic ancestry sites who appeared to be related to her had no relationship with Scholl, which began to raise some doubts in her mind. One woman's profile connected to Scholl's DNA test did not show any genetic links to Eve, which would have been the case if Scholl was the sperm donor for Margo.

Undaunted by the discovery, Eve explored other matches through DNA databanks of people seeking relatives and began corresponding with someone whose DNA match made her a genetic relative of Eve. When asked about her family, she disclosed to Eve that they lived in Texas. She identified one family member, with enough shared DNA to be a cousin, who identified her uncle as Dr. Kim McMorries, Margo's fertility doctor. A second male genetic cousin match also referred to his uncle, Dr. McMorries.

So, if Eve's genetic cousin was the niece of Dr. McMorries, then the riddle of Eve's genetic father was solved. The fertility doctor used his own sperm to fertilize Margo's egg. He was Eve's biological father. A leader in the field of genetic genealogy who consulted on the case, CeCe Moore, stated:

> In Eve's case, she was lucky. A lot of cousins and distant cousins of Dr. McMorries had taken DNA tests, and so we had a lot of people that we were able to compare against. I built out the doctor's family tree out to his great-great-great-grandparents because I want to see if Eve is carrying DNA of her mother's doctor's ancestors and there were plenty of them.

> I could see that she was carrying DNA from all the recent ancestral lines of Dr. McMorries' family, meaning all of his great-grandparents and of his grandparents. And so that gives me a lot of confidence that we've identified the correct biological father.

Eve's story raises a serious ethical issue of whether a doctor who performs *in vitro* fertilization with donor sperm should, without the consent of the woman and her husband seeking a pregnancy, use his own sperm, as admitted by Dr. McMorries.

Breaking the Pure Ethnic Descent

The final story can be found in novelist Dani Shapiro's captivating memoir titled *Inheritance*, in which she explores how a DNA ancestry test made her recalibrate her life and identity. Ms. Shapiro was brought up in what she calls "a large Orthodox Jewish clan." Her father died in an automobile accident when she was 23 years old, but he remained a strong presence in her life. Ms.

Shapiro was the only child of her father's second marriage, while her half-sister was from his first marriage.

When Ms. Shapiro took a DNA ancestry test, the results were puzzling, given what she was told about her ancestry. Throughout her life she learned that she was descended from generations of Orthodox Jews, traced to her third great-grandfather, from Eastern Europe in a Polish village. Her ancestry test results, however, reported 52 percent Eastern European Ashkenazi, with the remaining percentages being French, English, Irish, and German ancestry. This triggered the beginnings of her search for her identity, once the skepticism about the test results abated.

The first step in the journey was to compare her DNA profile to that of her half-sister, both performed by the same ancestry company. With 653,629 SNPs for comparison, it was concluded that they were not half-sisters – they were not even related. This left Ms. Shapiro with only one conclusion: either she or her presumed half-sister was not the biological child of their father. Digging deep into her past through living relatives, she learned that her parents had difficulty conceiving a child, which led them to a fertility clinic in Philadelphia. Without regulations, fertility clinics did not have to reveal the names of sperm donors. In addition, they could mix the sperm of young healthy donors with that of the husband to increase the odds of a pregnancy. Ms. Shapiro could not be certain about her paternity, since a paper trail of her birth did not exist. But the circumstances under which she apparently did not share any of her father's DNA made the sperm donor hypothesis the most likely scenario. The gnawing questions remaining were: Did her parents know there was donor DNA in her pregnancy? Who was her biological father? While the first question was never satisfactorily answered, the second question was resolved by other DNA ancestry tests.

Through a process of finding genetic relatives or at least clues to relatives on ancestry sites, and with the help of genealogical consultants, Ms. Shapiro was able to discover someone who had similar enough DNA to be her genetic cousin. That individual identified a person who was a doctor in her family. Eventually, with fearless persistence, Ms. Shapiro learned that her biological father was a sperm donor in Philadelphia and provided a sperm sample during the period her parents were treated at the fertility clinic. The stories

Ms. Shapiro was told throughout childhood about her ancestry of generations of observant Orthodox Jews was an adopted ancestry, and not her genetic ancestry, one that even her father may not have known.

And finally, a brief story of how ancestry testing brought an answer to an abandoned and adopted infant. Eleni Liff knew she was adopted as an infant. Her biological mother left her newborn in a shopping bag in the foyer of a Brooklyn apartment building. All Ms. Liff had was the note her mother had left in the bag: "May the angels watch over you and may God forgive me." Through a combination of DNA ancestry testing, sharing her genetic profile on public databases, and requesting her ancestry provider to check for close relatives, when Ms. Liff was 27 years old she was able to learn who her biological parents were.

These stories about finding unknown biological relatives and learning about falsely assumed biological relatives are the tip of the iceberg of such cases that would never have been possible without the use of genetics for assessing familial relations, particularly paternity.

CeCe Moore, a genetic genealogist and founder of DNA Detectives, reported that close to 10 percent of the people she consults will learn that the person who they believe is their father proves not to be their biological father.

Some of the same methods used in police investigations to identify family members through genetics are available to people exploring their family genetic links. Yet, there have been claims that the DNA ancestry tests are not as reliable as people make them out to be. The next chapter reviews the claims about the validity and reliability of DNA ancestry tests.

12 Accuracy, Consistency, and Validation of DNA Ancestry Tests

As a recreational activity, with no serious consequences at stake, it barely matters whether the results consumers receive from their DNA ancestry tests accurately represent the percentages of their ancestry from different geographical regions. Given that there are no international standards for such testing, unlike genetic disease tests, it is not surprising that the results from different ancestry testing companies vary. As noted in Chapter 4, there are several stages in the analysis of a person's saliva or cheek swabs where the criteria, reference frames, or analytics can vary among companies, yielding different outcomes.

Comparing Different Company Ancestry Results

Some people have sent their DNA samples to different companies to compare their results. Do the results agree and, if not, how great is the variation? Without a standard, a consumer can only learn about the variation of estimates across different companies. One could design a test to validate DNA ancestry results, as discussed in Chapter 4. Suppose we had evidence-based family genealogical trees for four or five generations, which means that each of the test participants had a full genealogical tree of their descent from their great-great-great-grandparents, where the genealogy was based on documents, not just memory or family lore. Then, each of those individuals completed a DNA ancestry application and test by a group of companies. Once the test results were in, we could compare each result with the individual's true descent ratios, based upon their documented family trees (e.g.,

60 percent Northern European, 30 percent Southeast Asian, 10 percent North African).

Ancestry companies make it a practice to estimate the confidence level of the test results or the potential error in assigning a particular region of descent. These are statistical error evaluations and are not based on comparing the results with people's actual ancestry. Their error analysis, which may provide a degree of confidence in their statistical methods, is not transparent to customers.

There have been a number of studies evaluating how effective panels of markers are in placing an individual to his/her ancestral population(s). One of the major sources of markers is the HapMap SNP (single nucleotide polymorphism) database developed by the US National Institutes of Health and deposited in the National Library of Medicine. The database contains 1,440,616 SNPs genotyped from 1,218 individuals, including Africans, Americans, Central South Asians, East Asians, and Europeans. Some results indicate that the more markers that are used, the better the results are for inferring ancestry, and that it is not a matter of selecting a small number of high-quality markers.

In a study by Pardo-Seco et al. of more than 400 AIMs (ancestry informative markers), the investigators predicted ancestry reasonably well, while those containing a few dozen AIMs show a large variability in ancestor estimates. Ironically, they found that a random selection of SNPs is equally as effective as a select panel when inferring ancestry in three continental populations: Europe, East Asia, and Africa. They discovered that 500–1,000 SNPs selected at random from the genome give an unbiased estimate of ancestry and perform as well as any AIM panel of similar size. One of their conclusions was that estimating genetic ancestry is strongly dependent on the number of AIMs used, and not on their individual informativeness.

Some DNA ancestry companies claim they used hundreds of thousands of markers to analyze a DNA sample. It is not clear whether their algorithms use a random subset of markers or whether the full set of markers (i.e., 800,000) provide a more precise and fine-grained biogeographical designation. Pardo-Seco et al. believe that for distinguishing ethnicities in very closely related populations one would have to use a genome-wide dataset of SNPs.

DNA Markers and Ancestry Regions

A study published in 2008 by Novembre et al. sought to determine whether the DNA markers of European subjects would cluster around certain regions. They began with a sample of 3,192 European individuals. For many of the sample subjects they were able to obtain the country of origin of the individual's grandparents, which was used to identify the geographical location that best represents the individual's ancestry. Where that information was unavailable, they used the self-reported country of birth. After the investigators eliminated individuals who failed to meet the study criteria, they had 1,387 individuals for whom they had high confidence of ancestry. Their analysis focused on 197,146 loci on the genomes of their participants. They analyzed the genomic data with an Affymetrix 500 K SNP chip.

The investigators used principal component analysis to map closely related genotypes from SNP information. They found that the cluster formations bore a notable resemblance to a geographical map of Europe. They concluded that with modern statistical techniques and sufficient markers, testing an individual's DNA can be used to infer their geographical origin with surprising accuracy – often to within a few hundred kilometers.

After learning about family inconsistencies in stories about descent, journalist Kirsten Brown wrote in 2018 that she turned to DNA tests. Brown signed up for DNA ancestry tests from several companies:

> I decided to conduct an experiment. I mailed my own spit samples to AncestryDNA, as well as to 23andMe and National Geographic. For each test I got back the story of my genetic heritage was different – in some cases wildly so … Four tests, four very different answers about where my DNA comes from – including some results that contradicted family history I felt confident was fact.

One company reported that Brown had 8 percent DNA from the Indian subcontinent, while another reported she had no South Asian DNA. She contacted the companies to learn why there were different results from the same DNA. One company responded that the estimates companies make are variable and depend on the methods they use, the alleles they choose to analyze, the reference panels they have, the algorithms they apply, and other customer samples they have in their databases. Brown concluded that it was more than the science that explained the different outcomes in her tests. Not only did the data matter, but also the interpretation of the data: "It's not that the science is bad. It's that it's inherently, an estimation based on how much our DNA matches up with people in other places around the world, in a world where people have been mixing and getting it on since the beginning of human history." Quoting one geneticist, Brown writes: "It's not that one's wrong and one's right. It's that there isn't an agreed-upon approach to pick the right number of markers and combine them mathematically. Everyone is sort of just making it up as they go along."

Most people who engage in direct-to-consumer DNA ancestry testing begin with inquisitiveness about what DNA can reveal about their ancestry. After receiving the test, they then begin to wonder whether the test is reliable and what another test from another company would reveal. If they do send out for more than one test, they enter a third stage of inquisitiveness: Why do the tests differ?

Another journalist, Rafi Letzter, writing for the science blog *Live Science* and acknowledging that he is Jewish and of Eastern European descent, with no doubts about his ancestry, skipped to the second stage. He took nine different commercial DNA ancestry tests under a variety of fictitious names, from three different companies, and compared the results. While the tests indicated some differences, they were at the margins. They all estimated that he was overwhelmingly Ashkenazi Jewish (85–100 percent of his DNA). Letzter sent back the same DNA under different names to one of those three companies and still received different results. This is a test of consistency. Citing a scientist he consulted, Letzter wrote: "the fact that they couldn't even produce consistent results from samples taken from the same person was a bit weird." Inconsistencies in results are sometimes premised on the assumption that there was one truth, albeit elusive. More likely, there are different ways of interpreting the data.

Ancestry Tests with Twins

Although the testing companies have different reference frames and use different algorithms for analyzing the data, one might expect consistency within their own testing regimes. A group of scientists at Case Western Reserve, MetroHealth Medical Center, and the Cleveland Clinic in Cleveland, Ohio evaluated the consistency of consumer genetic testing kits from three major DNA ancestry companies: 23andMe, AncestryDNA, and MyHeritage. For the DNA samples, the team recruited 21 twin-pairs (42 participants). They evaluated the concordance of ancestry results when twin-pairs were tested by the same company, and when twin-pairs were tested by different companies. Their results indicated that the concordance of twin genomes differed significantly between inter- and intra-company tests. When twin-pairs were tested by the same company, the mean percentage agreement ranged from 94.5 to 99.2 percent. The concordance of ancestry when participants were tested by different companies was lower, with the mean percentage ranging from 52.7 to 84.1 percent. The authors attributed the differences between inter- and intra-company testing consistency to the different reference databases used by different companies.

Submitting twin genomes to ancestry DNA companies was also carried out by a journalist twin who worked for the Canadian Broadcasting Company's (CBC) program *Marketplace*. Charlsie Agro of CBC and her twin sister bought ancestry DNA home kits from MyHeritage, 23andMe, and AncestryDNA, and compared the results. One company found differences in the identical twin's ancestry on Eastern European heritage (28 vs. 24.7), Italian (37 vs. 39), and Balkan (14 vs. 15). However, another company found that a majority of their heritage was Balkan (61 vs. 61) and Greek (20 vs. 19). On interviewing and quoting a population geneticist at one of the companies, Agro wrote: "Finding the boundaries is itself kind of frontiering science, so I would say that makes it kind of a science and an art."

That is the key point. There is science involved in DNA ancestry going back to allele frequencies and population genetics, but there are also discretionary assumptions made by different companies, such as the way different companies define the world's population regions, and the availability of proprietary data, which differ across companies.

Another interpretation for disparate results with twin ancestry DNA studies is the random quality of the analysis of long strands of DNA. A Stanford University website, Understanding Genetics, states that it is highly unlikely to get a similar result in the analysis of DNA samples. They can vary in a single individual from 1 to 20 percent fairly easily.

According to this reasoning, the computer algorithm splits up the sample DNA into thousands of segments, called "windows," which it analyzes one at a time. Each window may contain around 100 DNA markers. However, some windows cannot be read by the program. The Stanford site notes that when spots on the DNA cannot be read in a particular assay, that can tip the ancestry scales one way or another. In one read, a sequence of one identical twin's DNA is English, while the same sequence of DNA may be read as German for the second identical twin. A few differences of this nature can shift enough DNA to make the two not look identical from an ancestry point of view.

In other words, the computer program looks at the "windows" and if it finds one it cannot interpret, it examines those around it. If the "windows" to the left and right indicate German heritage, then it will conclude that the unin-terpreted "window" also signifies German heritage (Figure 12.1(a)).

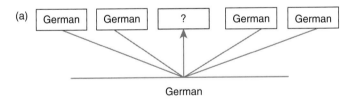

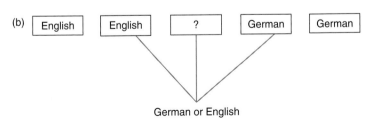

Figure 12.1   The series of windows analyzed by the algorithm for ancestry.

If the "windows" of the German and English heritage come side by side, the algorithm makes a choice (called smoothing). Sometimes it might decide the uninterpreted "window" is English and sometimes German, accounting for why twins can have different readings (Figure 12.1(b)). There are random operations in the computer algorithms that can give different results.

Is there a way to reduce the uncertainty so that twins would always be read the same? According to the Stanford analysis, you can get greater precision and consistency, but you would have to spend a lot more money on your test.

Self-Declared Ancestry vs. Genomic Ancestry

In a study by Lao et al., the authors evaluated the self-declared ancestry of US Americans, comparing their self-identity with the predicted genomic identity. Working with the genomic information from 664 US individuals, the researchers analyzed their SNPs in four main clusters: Asian, Hispanic, European, and African. Of those people who declared themselves European, 19 percent fell within the Hispanic cluster based on DNA analysis. Among those people who self-identified as being of African descent, a mere 2.2 percent were identified as Hispanic and 1.0 percent were identified as European from their genetic markers. Their self-identity of African ancestry was highly correlated with their genetic identity as being of African descent. Hispanics showed the most significant variations from self-identity and genomic identity (Table 12.1).

What is especially noteworthy about the methods of inferring ancestry from markers and reference populations is that the systems are self-correcting, or at least self-adjusting. The more data they have for ancestral populations, the more their programs are amended to take into consideration the new information. But the adjustments are proprietary, just as the reference populations and the statistical methods. There is also a certain degree of stochastic processes or randomness built into the system.

The American Society of Human Genetics convened a taskforce on ancestry testing. The recommendations, reported in 2010, called for standardization and guidelines:

Self-declared ancestry	Asian	Hispanics	European	African
US African	0%	2.2%	1.0%	96.8%
US European	0%	19.0%	80.6%	0.4%
US Hispanic	2.4%	77.8%	15.7%	4.0%
US Asian	99.9%	0.1%	0%	0%

Source: Lao et al. (2010), Evaluating self-declared ancestry of U.S. Americans with autosomal, Y-chromosomal and mitochondrial DNA. *Human Mutation*, 31: E1875–E1893. doi:10.1002/humu.21366.

Table 12.1   Results of the Lao et al. study

Academic researchers need to identify criteria for selecting the methods to be used, establish standards for statistical analysis, and create guidelines for peer review of proposals, and publications, and establish appropriate ways to convey findings to the general public. The ASHG suggests that there should be a "national roundtable discussion" of the sale of testing kits directly to the public.

After a decade, no progress has been made to meet the ASHG recommendations.

I shall end this chapter by quoting Pardo-Seco et al.: "Caution should be exercised when inferring ancestry using AIM panels. The concept of ancestry is a complex one and although it can be operational for particular purposes, it can lead to erroneous perceptions of human variability."

13 Conclusion

By this point in the book, we have explored multiple perspectives of DNA ancestry testing, beginning with its commercial success as a consumer recreational activity, its serious scientific foundations in population genetics, its applications in criminal investigations, and its social and ethical consequences in searching for one's identity.

The field of ancestry DNA testing is a work in progress. Companies continue to expand their population reference panels, refine their algorithms, and improve on the markers used in a sample to infer ancestry. Customer demand may be ahead of what companies can offer. People want to know the precise region in which their ancestors lived – the country, the city, the town. At various periods 3–6 generations ago and sporadically over thousands of years there has been considerable population movement. Regional boundaries were unclear, or if they were clear they often shifted after wars. There was considerable admixture of ethnic populations during mass immigrations.

The reference populations used today for ancestry testing consist of current inhabitants. They may assert, with little documented evidence, that their great-great-grandfather lived in the exact same region where they are living. As companies expand the reference populations, they may decrease the error rate or the bias in allele frequency in estimating ancestry. However, with larger populations they introduce more diversity, which adds some uncertainty, since with more alleles in the analysis there is a broader spread of the allele frequencies. This is analogous to the Heisenberg uncertainty principle in physics, where, regardless of how much information we have about the position and velocity of an elementary particle (i.e., an electron), the more certainty we have in measuring its position the less certainty we have in measuring its velocity.

The field of DNA ancestry testing would not have survived and thrived if most consumers were not satisfied with the results they receive from the companies that run the tests. As we saw in Chapter 11, some people who are at first skeptical of the results, on further investigation, learn that their lives were built on a deception, as their genetic profile proves to be correct.

Of course, there are many outliers, such as two monozygotic twins testing differently in their ancestry. Like any statistical sampling, if the test were repeated 100 times, maybe the results would converge to a single outcome with a standard deviation. The reference populations are not a random sample of the people living in a region. When the markers of a reference subject deviate too far from the cluster, the algorithm can remove that data point as an outlier.

The system of DNA ancestry inference is based on the differential frequency of markers, not a dozen, but thousands. When an SNP "P" has a frequency of 80 percent in population A, that means 8 out of 10 people in region A exhibit the SNP. In region B, the frequency of the SNP is 20 percent. If a person has SNP "P," then their ancestry is likely from region A. If the same or close percentages are repeated for thousands of SNPs, that provides strong evidence that the person's descent is from region A. But if the markers divide between regions, then the person's parents or grandparents may have come from two distinct regions. That is where mtDNA or Y chromosome DNA can resolve the issue.

A person who receives his/her ancestry test results and wants additional assurances of its accuracy may send it to several companies. If the results from several companies differ, which is very likely within certain ranges, the individual cannot know which test is more dependable since there is no universal standard. On the other hand, if several companies agree about the country of primary descent, and those results agree with what is known about one's family history, that may offer confidence that they have it right.

The population reference frames are another essential component of the DNA ancestry methodology. Writing in 2018, Cheung et al. warned consumers of the uncertainties, that despite ancestry declarations made by donors, many populations in these databases are admixed. They go on to say: "As commercial ancestry and phenotyping services become available, uncritical acceptance of predictions may result in misleading interpretations." There is much that can be learned from DNA ancestry, but the operative term is "caveat lector" – let the reader beware, there are areas of uncertainty.

Summary of Common Misunderstandings

DNA ancestry testing shows who your ancestors were. DNA ancestry testing compares your DNA markers to those of people living today in specific regions of the world (these are described as reference populations), and support inferences to your having ancestry from these regions under many assumptions. One of these assumptions is that all the ancestors of the people in the reference populations have always lived in the respective region, something that is difficult to document.

DNA ancestry testing can indicate your race/ethnicity. Whereas there exist DNA markers that are found more often in particular populations than in others (e.g., at the continental level), there are no markers that can indicate continental/racial/ethnic origin in any absolute sense. Rather, what the tests do is estimate the probability of being related to one or the other group, and this is what the respective ancestry proportions indicate. Most importantly, many people often receive results assigning them to more than one group, thus reflecting their mixed ancestry or to probabilities that assign them to two regions.

DNA ancestry testing can debunk what you knew about your ancestry and reveal your true ancestry. DNA ancestry testing is about genetic, not genealogical, ancestry. Because only half of a person's DNA is passed on to each of their offspring, the common DNA of an ancestor that is inherited by a descendant is smaller the greater the number of generations that separate them. As a result, only a small portion of a person's genealogical ancestors are reflected in that person's DNA, and so genetic ancestry reflects only a portion of a person's genealogical ancestry. Two individuals of the same genealogical ancestry can have no DNA in common.

DNA ancestry tests are useless because they cannot definitively reveal your ancestry. DNA ancestry tests indeed cannot definitively reveal a person's ancestry within a range of hundreds of years. However, they can identify close or distant relatives with a high degree of accuracy. Therefore, they can be very useful to people who do not know much about their ancestry and so look for their parents or other relatives.

DNA ancestry testing is consistent within and among companies. The different companies use different algorithms, statistical programs, reference databases, and genetic markers to infer ancestry. As a result, people who have taken tests from different companies have received different results. Furthermore, as data from more people is added to the databases, and the reference populations that the companies use for comparison change over time, the results for the same person can also change over time.

Taking a DNA ancestry test concerns the test-taker only. This is not the case because the test-taker's relatives might also be affected when their results are uploaded in the respective databases. Because these tests can document identity by descent, it is possible for people to be contacted by other people that the DNA data has indicated as being close or distant relatives. Finally, as in the case of the Golden State Killer, it is possible to find a person even if they have never taken a test. If the DNA of a person is available – for instance, from a crime scene – it is possible to use it to find people who are related to the suspect through the databases. Then, once such matches are found, other kinds of data (genealogical, etc.) can be used to identify the suspect.

References

Chapter 1

On the biogeography of ancient human populations:

Luca Cavalli-Sforza, L., Menozzi, P., and Piazza, A. *The History and Geography of Human Genes*. Princeton, NJ: Princeton University Press, 1994.

On how we understand "ancestry":

Mathieson, I., and Scally, A. What is ancestry? *PLoS Genetics* 2020. doi: 10.1371/journal.pgen.1008624. Rak, J. Radical connections: genealogy, small lives, big data. *Journal of Auto/Biography Studies* 32(3):479–497 (2017). TallBear, K. *Native American DNA: Tribal Belonging and the False Promise of Genetic Science*. Minneapolis, MN: University of Minnesota Press, 2013.

Chapter 2

On the growth of the ancestry DNA consumer sector:

Copeland, L. Who was she? A DNA test revealed this mystery. *Washington Post* July 30, 2017. Bursztynsky, J. *Health Tech Matters*, CNBC. February 12, 2019. www.cnbc.com/2019/02/12/privacy-concerns-rise-as-26-million-share-dna-with-ancestry-firms.html; Royal, C.D., Novembre, J., Fullerton, S.M., et al. Inferring genetic ancestry: opportunities, challenges and implications. *American Journal of Human Genetics* 86:661–673 (2010). Wagner, J.K., Cooper, J.D., Sterling, R., and Royal, C.D. Tilting at windmills no longer: a data-driven discussion of DTC DNA tests. *Genetics in Medicine* 14

(6):586–593 (2012); Watt, E. and Kowal, E. To be or not to be Indigenous? Understanding the rise of Australia's Indigenous population since 1971. *Ethnic and Racial Studies* 42(16): 63–82 (2019).

On the dual business model (two-sided market) of ancestry companies:

Phillips, A.M. Only a click away: DTC genetics for ancestry, health, love . . . and more – a view of the business and regulatory landscape. *Applied & Translational Genomics* 8:16–22 (2016). Versel, N. Ancestry grant marks new direction for Wolters Kluwer's UpToDate. GenomeWeb. December 5, 2019. Stoekle, H.-C., Manzer-Bruneel, M.-F., Vogt, G., and Hervé, C. 23andMe: a new two-sided data-banking market model. *BMC Medical Ethics* 17:19–27 (2016). Defrancesco, L., and Klevocz, A. Your DNA broker. *Nature Biotechnology* 37:842–847 (2009).

On criticism of ancestry testing:

Owermohle, S. As genetic testing companies look to enter the medical arena, criticism grows. SNL Kagan Media & Communications Report. Charlottesville, May 25, 2018; Hamilton, A., Invention of the year. *Time.* http://content.time.com/time/specials/packages/article/0,28804,1852747_1854493_1854113,00.html; Pollack, A. F.D.A. orders genetic testing firm to stop selling DNA analysis service. *New York Times*, November 26, 2013; www.cnet.com/news/23andme-fda-approval-home-dna-test-genetic-risk; Watt, E., and Kowal, E. What's at stake? Determining indigeneity in the era of DTY DNA. *New Genetics & Society* 38(2):142–164 (2019); Soo-Jim Lee, S., Bolnick, D.A., Duster, T., Ossario, P., and TallBear, K. The illusive gold standard in genetic ancestry testing. *Science* 325:38–39 (2009).

On the second wave of DNA ancestry testing:

Callaway, Z. Ancestry testing goes for pinpoint accuracy. *Nature* 486:17 (2012); Globe News Wire. Ancestry.Com launches new ancestry DNA service. May 3, 2012; Regaldo, A. More than 26 million people have taken an at home ancestry test. *Technology Review*, February 11, 2019; https://blogs.ancestrycom/ancestry/2019/05/31/ancestry-surpasses-15-million-dna-customers; Strong, C. A., Martin, B.A.S., Jin, H.S., Greer, D., and O'Connor, P. Why do consumers research their ancestry? Do self-uncertainty and the need for closure influence consumer's involvement in ancestral products? *Journal of Business Research* 99:332–337 (2019); Creet, J. Home genealogy kit sales plummet over data privacy concerns. *The Conversation*, March 1, 2020; https://theconversation .com/home-genealogy-kit-sales-plummet-over-data-privacy-concerns-132082.

On biogeographical ancestry:

Gannett, L. Biogeographical ancestry and race. *Studies in the History and Philosophy of Biological and Biomedical Sciences* 47:173–184 (2014).

Chapter 3

On ancient human migrations:

Cavalli-Sforza, L.L., and Cavalli-Sforza, F. *The Great Human Diaspora*. Reading, MA: Perseus Books, 1995; Lipson, M., Ribot, I., Mallick, S., et al. Ancient West African foragers in the context of African population history. *Nature* 577:665–670 (2020); Luca Cavalli-Sforza, L., Menozzi, P., and Piazza, A. *The History and Geography of Human Genes*. Princeton, NJ: Princeton University Press, 1994. Rosenberg, N.A., Pritchard, J.K., Weber, J.L., et al. Genetic structure of human populations. *Science* 298:2381–2384 (2002); Wood, B. Evolution: origin(s) of modern humans. *Evolutionary Biology* 27:R746–R769 (2017).

On human genetic diversity:

Lewontin, R.C. The apportionment of human diversity. *Evolutionary Biology* 6:381 (1972). Hunley, K.L., Cabana, G.S., and Long, J.C. The apportionment of human diversity revisited. *American Journal of Physical Anthropology* 160:561–569 (2016); Choudhury, A., Aron, S., Songupta, D., Hazelhurst, S., and Ramsay, M. African genetic diversity provides novel insights into evolutionary history and local adaptations. *Human Molecular Genetics* 27(r2):209–218 (2018); Lewontin, R. The apportionment of human diversity. In: *Evolutionary Biology*, T. Dobzhansky, M.K. Hech, W.C. Steer, eds. New York: Springer, 1972, pp. 381–398; Edwards, A.W.F., Campbell, M.C., and Tishkoff, S.A. African genetic diversity: implications for human demographic history, modern human origins, and complex disease mapping. *Annual Review of Genomics & Human Genetics* 9:403–433 (2008); Barbujani, G., Magagni, A., Minch, E., and Cavalli-Sforza, L.L. An apportionment of human DNA diversity. *Proceedings of the National Academy of Sciences of the USA* 94:4516–4519 (1997). National Institute of Health, Human Genome Research Institute. www.genome.gov/about-genomics/fact-sheets/Genetics-vs-Genomics#:~:text=All%20human%20beings%20are%2099.9,about%20the%20causes%20of%20diseases.

On genetic structure of human populations:

Rosenberg, N.A., Prichard, J.K., Weber, J.L., et al. Genetic structure of human populations. *Science* 298:2381–2385 (2002).

On genetics and Native American ancestry:

TallBear, K. *Native American DNA: Tribal Belonging and the False Promise of Genetic Science*. Minneapolis, MN: University of Minnesota Press, 2013.

Chapter 4

On selecting and evaluating ancestry markers:

Rosenberg, N. Algorithms for selecting informative marker panels for population assignment. *Journal of Computational Biology* 12(9):1181–1201 (2005); Joshua Sampson, K.K., Kidd, J.R., and Kidd, H.Z. Selecting SNPs to identify ancestry. *Annals of Human Genetics* 75(4):539–553 (2011); Rosenberg, N., Prichard, J.K., Weber, J.L., et al. Genetic structure of human populations. *Science* 298:2381–2385 (2002); Rosenberg, N.A., Li, L.M., Ward, R., and Pritchard, J.K. Informativeness of genetic markers for inference of ancestry. *American Journal of Human Genetics* 73:1402–1422 (2003); Pardo-Seco, J., Martinón-Torres, F., and Salas, A. Evaluating the accuracy of AIM panels at quantifying genome ancestry. *BMC Genomics* 15:543–555 (2014); Al-Asfi, M., McNevin, D., Mehta, B., et al. Assessment of the precision ID ancestry panel. *International Journal of Legal Medicine* 132:1581–1594 (2018); Gettings, K.B., Lai, R., Johnson, J. L., et al. A 50-SNP assay for biogeographic ancestry and phenotype prediction in the U.S. population. *Forensic Science International: Genetics* 8:101–108 (2014); Halder, I., Shriver, M., Thomas, M., Fernandez, J.R, and Frudakis, T. A panel of ancestry informative markers for estimating individual biogeographical ancestry and admixture from four continents: utility and applications. *Human Mutation* 29(5):648–658 (2008); Jin, X.-Y.,

Cui, W., Chong, C., et al. Biogeographic origin prediction of three continental populations through 42 ancestry informative SNPs. *Electrophoresis* 41:235–245 (2019); Paschou, P., Ziv, E., Burchard, E.G., et al. PCA-correlated SNPs for structure identification in worldwide human populations. *PLoS Genetics* 3(9):1–15 (2007); Soundararajan, U., Yun, L., Shi, M., and Kidd, K.K. Minimal SNP overlap among multiple panels of ancestry informative markers argues for more international collaboration. *Forensic Science International: Genetics* 23:25–32 (2016); Kidd, K.K., Speed, W.C., Pakstis, A.J., et al. Progress toward an efficient panel of SNPs for ancestry inference. *Forensic Science International: Genetics* 10:23–32 (2014); Frudakis, T., Vonkateswarlu, K., Thomas, M.J., et al. A classifier for the SNP-based inference of ancestry. *Journal of Forensic Science* 48(4):1–8 (2003); Hwa, H.-L., Lin, C.-P., Huang, T.-Y., et al. A panel of 130 autosomal single-nucleotide polymorphisms for ancestry assignment in five Asian populations and in Caucasians. *Forensic Science Medical Pathology* 13:177–187 (2017); Gao, X. and Starmer, J. Human population structure detection via multilocus genotype clustering. *BMC Genetics* 8:34–45 (2007); Kersbergen, P., van Duijn, K., Kloosterman, A.D., et al. Developing a set of ancestry-sensitive DNA markers reflecting continental origins of humans. *BMC Genetics* 10:69–72 (2009); Patterson, N., Price, A. L., and Reich, D. Population structure and eigen analysis. *PLoS Genetics* 2 (12):2074–2093 (2006).

On patent application for ancestry DNA tests:

US Patent Application. US200410229231A. Compositions and methods for inferring ancestry. November 18, 2004. Inventors: Frudakis, T.N., and Shriver, M.D. Abandoned December 8, 2013.

On genetic diversity within and between populations:

Witherspoon, D.J., Wooding, S., Rogers, A.R., et al. Genetic similarities within and between human populations. *Genetics* 176:357–359 (2007).

On validating DNA ancestry markers and inference:

Chaitanya, L., van Oven, M., Weiler, N., et al. Developmental validation of mitochondrial DNA genotyping assays for adept matrilineal inference of biogeographic ancestry at a continental level. *Forensic Science International: Genetics* 11:39–51 (2014); Royal, C.D., Novembre, J., Fullerton, S.M., et al. Inferring genetic ancestry: opportunities, challenges, and implications. *American Journal of Human Genetics* 86:661–673 (2010); Nassir, R., Kosoy, R., Tian, C., et al. An ancestry informative marker set for determining continental origin: validation and extension using human genome diversity panels. *BMC Genetics* 10:39–52 (2009); Pardo-Seco, J., Martinón-Torres, F., and Salas, A. Evaluating the accuracy of AIM panels at quantifying genome ancestry. *BMC Genomics* 15:543–555 (2014); Chaitanya, L., van Oven, M., Weiler, N., et al. Developmental validation of mitochondrial DNA genotyping assays for adept matrilineal inference of biogeographic ancestry at a continental level. *Forensic Science International: Genetics* 11:39–51 (2014).

On sources of human genomic data:

Wells, S. *Deep Ancestry: Inside the Genographic Project*. Washington, DC: National Geographic, 2006.

On ancestry company information:

Ancestry.com, Ethnicity estimate 2018 white paper; www.ancestrycdn.com/dna/ static/images/ethnicity/help/WhitePaper_Final_091118dbs.pdf.

Chapter 5

On ancestry informative markers:

Frudakis, T., Venkateswartu, K., Thomas, M.J., et al. A classifier for the SNP-based inference of ancestry. *Journal of Forensic Science* 48(4):1–14 (2003).

National Human Genome Research Institute www.genome.gov/genetics-glossary /Ancestry-informative-Markers; Kersbergen, P., van Duijn, K., Kloosterman, A. D., et al. Developing a set of ancestry-sensitive DNA markers reflecting continental origins of humans. *BMC Genetics* 10:69–72 (2009); Chakraborty, R., Kamboh, M.L., and Ferrell, R.E. "Unique" alleles in admixed populations: a strategy for determining "hereditary" population differences. *Ethnicity & Disease* 1:245–256 (1991); Neel, J.V. "Private" genetic variants and the frequency of mutation among South American Indians. *Proceedings of the National Academy of Sciences* 70(12):3311–3315 (1973); Halder, I.,

Shriver, M., Thomas, M., Fernandez, J.R., and Frudakis, T. A panel of ancestry informative markers for estimating individual biogeographical ancestry and admixture from four continents: utility and applications. *Human Mutation* 29 (5):648–658 (2008); Royal, C.D., Novembre, J., Fullerton, S.M., et al. Inferring genetic ancestry: opportunities, challenges and implications. *American Journal of Human Genetics* 86:661–673 (2010); Paschou, P., Lewis, J., Javed, A., et al. Ancestry informative markers for fine-scale individual assignment to worldwide populations. *Journal of Medical Genetics* 47(12):835–847 (2010); Soundararajan, U., Yun, L., Shi, M., and Kidd, K.K. Minimal SNP overlap among multiple panels of ancestry informative markers argues for more international collaboration. *Forensic Science International: Genetics* 23:25–32 (2016).

On mitochondrial ancestry:

Kivisild, T. Maternal ancestry and population history from whole mitochondrial genomes. *Investigative Genetics* 6:3–13 (2015).

On PCR amplification:

Roewer, L., and Epplen, J.T. Rapid and sensitive typing of forensic stains by PCR amplification of polymorphic simple repeat sequences in case work. *Forensic Science International* 53(2):163–171 (1992).

On Y chromosome DNA:

Kayser, M. Forensic use of Y-chromosome DNA: a general overview. *Human Genetics* 136:621–635 (2017).

On reading your ethnicity:

Ancestry.com. More than a pie chart and a number: reading your ethnicity estimate; www.ancestry.com/lp/ethnicity-estimate/reading-your-ethnicity-estimate.

On the accuracy of AIMs:

Pardo-Seco, J., Martinón-Torres, F., and Salas, A. Evaluating the accuracy of AIM panels at quantifying genome ancestry. *BMC Genetics* 14:543–555 (2014);

Nassir, R., Kosoy, R., Tian, C., et al. An ancestry informative marker set for determining continental origin: validation and extension using human genome diversity panels. *BMC Genetics* 10:39–52 (2009).

On admixture and ancestry:

Galanter, J.M., Fernandez-Lopez, J.C., Gignoux, C.R., et al. Development of a panel of genome-wide ancestry informative markers to study admixture throughout the Americas. *PLoS Genetics* 8(3):1–15 (2012).

On forensic analysis of DNA:

FBI. Combined DNA Index System (CODIS); www.fbi.gov/services/laboratory/biometric-analysis/codis.

Chapter 6

On reference panels of human genomic data:

Santos, C., Phillips, C., Fabro, O., et al. Completion of a worldwide reference panel of samples for an ancestry indel assay. *Forensic Science International: Genetics* 17:75–80 (2015); Yale Center for Medical Informatics. The Allele Frequency Database; https://alfred.med.yale.edu/alfred/ALFREDtour-overview.asp; Pakstis, A.J., Kang, L., Liu, L., et al. Increasing the reference populations for the 55 AISNP panel: the need and benefits. *International Journal of Legal Medicine* 131:913–917 (2017); Lee, J.H., Cho, S., Kim, M.-Y., et al. Genetic resolution of applied biosystems precision ID ancestry panel for seven Asian populations. *Legal Medicine* 34:42–47 (2018); Cann, H.M, de Toma, C., Cazes, L., et al. A human diversity cell line. *Science* 296 (5566):261–262 (2002); Nelson, M.R., Bryc, K., King, K.S., et al. The population reference sample, POPRES: a resource for population, disease, and pharmacological genetics research. *American Journal of Human Genetics* 83:347–358 (2008).

On the accuracy of AIMs panels:

Pardo-Seco, J., Martinón-Torres, F., and Salas, A. Evaluating the accuracy of AIM panels at quantifying genome ancestry. *BMC Genomics* 15:543 (2014).

On AncestryDNA white paper:

AncestryDNA. Ethnicity estimate 2019 white paper; www.ancestry.com/dna/r esource/whitePaper/AncestryDNA-Ethnicity-White-Paper.pdf; AncestryDNA. Ethnicity estimate 2018 white paper; www.ancestrycdn.com/dna/static/pdf/ whitepapers/EV2019_white_paper_2.pdf. 23andMe. Ancestry composition. www.23andme.com/en-gb/ancestry-composition-guide.

On human genetic variation:

McVean, G.A., Abecasis, D.M., Auton, A., et al. An integrated map of genetic variation from 1,092 human genomes. *Nature* 491(7422):56–65 (2012); Sankar, P., and Cho, M.K. Genetics: toward a new vocabulary of human genetic variation. *Science* 298(5597):1337–1338 (2002).

Chapter 7

On population structure and ancestry:

Shriver, M.D., Smith, M.W., Jin, L., et al. Ethnic-affiliation estimation by use of population- specific DNA markers. *American Journal of Human Genetics* 60:957–964 (1997); Rosenberg, N.A. A population-genetic perspective on the similarities and differences among worldwide human populations. *Human Biology* 83(6):659–684 (2011); Wright, S. The general structure of populations: Galton lecture at University College, London. *Annals of Eugenics* 15:323–354 (1951).

On 23andME ancestry company:

23andMe. Ancestry composition. www.23andme.com/en-gb/ancestry-composition-guide.

On F_{ST} statistics and ancestry:

Wright, S. The interpretation of population structure by F-statistics with special regard to systems of mating. *Evolution* 19:395–420 (1965); Holsinger, K.E., and Weir, B.S. Genetics in geographically structured populations: defining, estimating and interpreting F_{ST}. *Nature Review of Genetics* 10(9):639–650 (2009); Jakobsson, M., Edge, D., and Rosenberg, N.A. The relationship between F_{ST} and the frequency of the most frequent alleles. *Genetics* 193:515–528 (2013).

Chapter 8

On AncestryDNA White Paper:

AncestryDNA. Ethnicity estimate 2019 white paper.; www.ancestry.com/dna/res ource/whitePaper/AncestryDNA-Ethnicity-White-Paper.pdf;

On patents for DNA probe:

US Patent. Target analyte sensors utilizing microspheres. Patent No. 6,859,570. Inventors Walt, D.R., and Michael, K.L. February 22, 2005. Assignee: Trustees of Tufts College, Tufts University; U.S. Patent Composite arrays utilizing microspheres. Patent No. 6,858,394. Issued February 22, 2005. Filed December 28, 1999. Assignee Illumina Inc. Inventors: Chee, M.S., Auger, S. R., and Stuelpnagel, J.R.; U.S. Patent Array compositions for improved signal detection. Patent No. 6,942,698. Inventors: Dickenson, T.A., Meade, S., Bernard, S.M., et al. Filed August 30, 2000.

On microarrays:

Grunstein, M., and Hogness, D.S. Colony hybridization: a method for the isolation of cloned DNAs that contain a specific gene. *Proceedings of the National Academies of Sciences, USA* 72:3961–3965 (1975); Wikipedia. Microarray; https://en.wikipedia.org/wiki/microarray; Chang, T.-W. Binding of cells to matrixes of distinct antibodies coated on solid surface. *Journal of Immunological Methods* 65:217–223 (1983); Interstate Technology Regulatory Council (ITRC), Environmental Molecular Diagnostics (EMD) Team. Fact sheet on microarrays; www.itrcweb.org/do cuments/team_emd/microarrays_fact_sheet.pdf; Ferguson, J.A., Christian Boles, T., Adams, C.P., and Walt, D.R. A fiber-optic DNA biosensor micro-array for the analysis of gene expression. *Nature Biotechnology* 14:1681–1684 (1996).

Chapter 9

On the history of forensic DNA identification:

Krimsky, S., and Simoncelli, T. *Genetic Justice: DNA Databanks, Criminal Investigation and Civil Liberties*. New York: Columbia University Press, 2011;

FBI. Combined DNA Index System (CODIS); www.fbi.gov/services/laboratory/biometric-analysis/codis.

On the Golden State Killer:

Dery, G.M., III. Can a distant relative allow the government access to your DNA? The Fourth Amendment implications of law enforcement's genealogical search for the Golden State killer and other genetic genealogy investigations. *Hastings Science & Technology Law Journal* 10:103 (2019); Greytak, E.M., Moore, C.C., and Armentrout, S.L. Genetic genealogy for cold case and active investigations. *Forensic Science International* 299:103–113 (2019).

On open-source DNA databases:

Verogen. A message to Verogen customers about the GEDmatch partnership; https://verogen.com/a-message-to-verogen-customers-about-the-gedmatch-partnership;
Guest, C. DNA and law enforcement: how the use of open-source DNA databases violates privacy rights. *American University Law Review* 68:1015 (2019); Hodge, S.D., Jr. Current controversies in the use of DNA in forensic investigations. *University of Baltimore Law Review* 48:39–66 (2018).

On phenotyping:

Frudakis, T.N. *Molecular Photofitting: Predicting Ancestry and Phenotype Using DNA*. Burlington, MA: Academic Press, 2008; Wolinsky, H. CSI on steroids. *EMBO Reports* 16(7):782–786 (2015); Fausto-Sterling, A. *Boston Review: Cold Case*, blog. August 11, 2015; www.AnneFaustoSterling.com/Boston-review-cold-case.

Chapter 10

On ownership of personal cells:

Rebecca, S. *The Immortal Life of Henrietta Lacks*. New York, Random House, 2011; Krimsky, S. *Science in the Private Interest*. Lanham, MD: Rowman & Littlefield, 2003.

On privacy of one's DNA:

Erlich, Y., Shor, T., Pe'er, I., and Carmi, S. Identity inference of genomic data using long-range familial searches. *Science* 362(6415):690–694 (2018); Wallace, S. E., Gourna, E.G., Nikolova, V., and Sheehan, N.A. Family tree and ancestry inference: is there a need for "generational" consent. *BMC Medical Ethics* 16:87–96 (2015); Brown, K.V. deleting your online DNA data is brutally difficult: a reporter's effort to erase her genetic footprint gets snared in a thicket of policies and rules. *Bloomberg News*, June 15, 2018; www.bloomberg.com/news/articles/2018–06–15/deleting-your-online-dna-data-is-brutally-difficult; Sklar, E.B. Be careful where you spit: do HIPAA-covered genetic tests actually provide privacy protection to consumers? *Seton Hall Legislative Journal* 44:177 (2020); Kody, H.A. Standing to challenge familial searches of commercial DNA databases. *William and Mary Law Review* 61:287–318 (2019).

On genetic ancestry testing:

Soo-Jin Lee, S., Bolnick, D.A., Duster, T., Ossorio, P., and TallBear, K. The illusive gold standard in genetic ancestry testing. *Science* 325:38–39 (2009); Sarata, A. K. Congressional research services: genetic ancestry testing. March 12, 2008; Regalado, A. 2017 was the year consumer DNA testing blew up. *MIT Technology Review*, February 12, 2018.

On forensic DNA searches:

Edge, M., and Coop, G. How lucky was the genetic investigation in the Golden State Killer case? bioRXiv, January 28, 2019; Greytak, E.M., Moore, C.C., and Armentrout, S.L. Genetic genealogy for cold case and active investigations. *Forensic Science International* 299:103–113 (2019); Zabel, J. The killer inside us: law, ethics and the forensic use of family genetics. *Berkeley Journal of Criminology* 24:47–100 (2019); Moreau, Y. Crack down on genomic surveillance. *Nature* 576:36–38 (2019).

On genetic identity:

Tutton, R. "They want to know where they came from": population genetics, identity and family genealogy. *New Genetics and Society* 23(1):105–120 (2004); Korthase, A. Seminal choices: the definition of "Indian Child" in a time of assisted reproduction technology. *Journal of the American Academy of Matrimonial Law* 31:131–156 (2018); Zimmer, C. Elizabeth Warren has a Native American ancestor: does that make her Native American? *New York Times*, October 15, 2018; Walajahi, H., Wilson, D.R., and Chandros Hull, S. Constructing identities: the implications of DTC ancestry testing for tribal communities. *Genetic Medicine* 21(8):1744–1750 (2019); Scully, M., Brown, S.D., and King, T. Becoming a Viking: DNA testing, genetic ancestry and placeholder identity. *Ethnic and Racial Studies* 39(2):162–180 (2016).

On race and genetics:

Jobling, M.A., Rasteiro, R., and Wetton, J.H. In the blood: the myth and reality of genetic markers of identity. *Ethnic and Racial Studies* 39(2):142–161 (2016); Nelson, A. *The Social Life of DNA: Race, Reparation and Reconciliation after the Genome*. Boston, MA: Beacon Press, 2016; Bernstein, D. May an individual claim minority status based on a DNA test showing a small amount of African heritage? *Reason Magazine*, January 8, 2020;

Orion Insurance Group *v.* Washington's Office of Minority and Women's Business Enterprises et al. United States Court of Appeals for the 9th Circuit. No. 17-35749. D.C. No. 3:16-cv-05582-RJB. Appeal from the U.S. District Court for Western District of Washington. Argued December 3, 2018; Kowal, E., and Llamas, B. Race in a genome: long read sequencing, ethnically-specific reference genomes and the shifting horizon of race. *Journal of Anthropological Sciences* 97:91–106 (2019); Roth, W.D., Yaglici, S., Jaffe, K., and Richardson, L. Do genetic ancestry tests increase racial essentialism? Findings from a randomized controlled trial. *PLoS One* 2020, doi: 10.1371/journal.pone.0227399; Duster, T. Race and reification in science. *Science* 307:1050–1052 (2005); Duster, T. Ancestry testing and DNA. In: *Race and the Genetic Revolution*. Sheldon Krimsky and Kathleen Sloan, eds. New York: Columbia University Press, 2011; Reich, D. How genetics is changing our understanding of "race." *New York Times*, March 23, 2018; Reich, D. How to talk about "race" and genetics. *New York Times*, March 30, 2018; Horowitz, A. I., Saperstein, A., Little, J., Maiers, M., and Hollenbach, J.A. Consumer (dis-)interest in genetic testing: the roles of race, immigration, and ancestral certainty. *New Genetics and Society* 38(2):165–194 (2019).

On genetics and ethnicity:

Schwartz, O. What does it mean to be genetically Jewish? DNA tests have been used in Israel to verify a person's Jewishness. *Guardian*, June 13, 2019; Roth, W.D., and Ivemark, B. Genetic options: the impact of genetic ancestry testing on consumers' racial and ethnic identities. *American Journal of Sociology* 124(1):150–184 (2018); Bryc, K., Durand, E.Y., McPherson, M., et al. The genetic ancestry of African Americans, Latinos and European Americans across the United States. *American Journal of Human Genetics* 96:37–53 (2015); Hochschild, J.L., and Sen, M. To test or not? Singular or multiple heritage? *Dubois Review* 12(2):321–347 (2015); Elliott, C., and Brodwin, C. Identity and genetic ancestry tracing. *British Medical Journal* 325 (7378):1469–1471 (2002).

Chapter 11

On unexpected paternity:

Lee, C.L. Current status of paternity testing. *Family Law Quarterly* 9:615–633 (1975); DNA Diagnostic Center. History of DNA testing; http://dnacenter .com/history-dna-testing; Schreiber, S.L. Truth: a love story. *Harvard Magazine*, July–August 2019; Copeland, L. A DNA test revealed a mystery. *Toronto Star*, August 7, 2017; Cape Breton Post. Trying to unknot a long-held family secret: "I would like to understand my genetic makeup!" July 27, 2019; www.capebretonpost.com/lifestyles/trying-to-unknot-a-long-held-family-secr et-i-would-like-to-understand-my-genetic-makeup–327248. ABC. *20/20. Seed of Doubt; Eve Wiley's Story*. 2019; Shapiro, D. *Inheritance: A Memoir of Genealogy, Paternity and Love*. New York: Anchor Books, 2019.

On parental discovery through DNA ancestry:

Kilgannon, C. She was left in a bag as a newborn: DNA testing helped her understand why. *New York Times*, May 22, 2019; www.nytimes.com/2019/0 5/22/nyregion/baby-abandoned-birth-paretns.html; Erica, E. Is DNA testing telling us more than we want to know? The untold story of Ancestry.com. *Desert News*, May 30, 2018; www.deseret.com/2018/5/31/20644896/is-dna-testing-telling-us-more-than-we want-to-know-the untold-story-of-ancestry-com.

Chapter 12

On ancestry predictions:

Jini, X.-Y., Cui, W., Chen, C., et al. Biogeographic origin predictions of three continental populations through 42 ancestry informative SNPs. *Electrophoresis* 41:235–245 (2020); Novembre, J., Johnson, T., Bryc, K., et al. Genes mirror geography within Europe. *Nature* 456(6):98–103 (2008).

On the accuracy of ancestry tests:

Pardo-Seco, J., Martinón-Torres, F., and Salas, A. Evaluating the accuracy of AIM panels at quantifying genome ancestry. *BMC Genomics* 15:543–555 (2014); Brown, K. How DNA testing botched my family's heritage, and probably yours, too. *Gizmodo*, January 16, 2018; https://gizmodo.com/how-dna-testing-botched-my-familys-heritage-and-probab-1820932637; Letzter, R. I took 9 different commercial DNA tests and got 6 different results. *Live Science*, November 5, 2019; Huml, A.M., Sullivan, C., Figueroa, M., Scott, K., and Sehgal, A.R. Consistency of direct-to-consumer genetic testing results among identical twins. *American Journal of Medicine* 133(1):143–146 (2020); Agro, C., and Denne, L. Twins get some "mystifying" results when they put 5 DNA ancestry kits to the test. *CBC Technology & Science Marketplace*, January 18, 2019; www.cbc.ca/news/technology/dna-ancestry-kits-twins-marketplace-1.4980976; The Tech Interactive. Understanding genetics; https://genetics.thetech.org/ask-a-geneticist/same-dna-different-ancestry-results; American Society of Human Genetics. Report of ancestry taskforce. *Biotechnology Law Report* 4:389 (2010).

On the HapMap database:

NCBI. NCBI retiring HapMap resource; http://hapmap.ncbi.nlm.nih.gov/

Chapter 13

On the uncertainties and gains of DNA ancestry testing:

Sankar, P., and Cho, M.K. Genetics: toward a new vocabulary of human genetic variation. *Science* 298 (5597):1337–1338 (2002). Cheung, E.Y.Y., Gahan, M. E., and McNevin, D. Predictive DNA analysis for biogeographical ancestry. *Australian Journal of Forensic Sciences* 50(6):651–658 (2018).

Index

Tables are referred to in **bold**, figures in *italics*

Other books authored, coauthored, or coedited by Sheldon Krimsky

Genetic Alchemy: The Social History of the Recombinant DNA Controversy (1982). The MIT Press.

Environmental Hazards: Communicating Risks as a Social Process (1988). Auburn House.

Biotechnics and Society: The Rise of Industrial Genetics (1991). Praeger.

Social Theories of Risk (1992). Praeger.

Agricultural Biotechnology and the Environment (1996). University of Illinois Press.

Hormonal Chaos: The Scientific and Social Origins of the Environmental Endocrine Hypothesis (2000). Johns Hopkins University Press.

Science and the Private Interest (2003). Rowman & Littlefield.

Rights and Liberties in the Biotech Age (2005). Rowman & Littlefield.

Genetic Justice: DNA Databanks, Criminal Justice and Civil ‧Liberties (2011). Columbia University Press.

Race and the Genetic Revolution: Science, Myth and Culture (2011). Columbia University Press.

Biotechnology in Our Lives (2013). Skyhorse Publishing.

Genetic Explanation: Sense & Nonsense. (2013). Harvard University Press.

The GMO Deception (2014). Skyhorse Publishing.

Stem Cell Dialogues: A Philosophical and Scientific Inquiry into Medical Frontiers (2015). Columbia University Press.

Conflict of Interest in Science (2018). Skyhorse Publishing.

GMOs Decoded (2019). The MIT Press.